William Henry Searles

The Railroad Spiral

The Theory

William Henry Searles

The Railroad Spiral
The Theory

ISBN/EAN: 9783744678452

Printed in Europe, USA, Canada, Australia, Japan

Cover: Foto ©berggeist007 / pixelio.de

More available books at **www.hansebooks.com**

THE

RAILROAD SPIRAL.

THE THEORY OF THE

COMPOUND TRANSITION CURVE

REDUCED TO

PRACTICAL FORMULÆ AND RULES FOR APPLICATION IN FIELD WORK;

WITH

COMPLETE TABLES OF DEFLECTIONS AND ORDINATES FOR FIVE HUNDRED SPIRALS.

BY

WILLIAM H. SEARLES, C.E.,

MEMBER AMERICAN SOCIETY OF CIVIL ENGINEERS,
AUTHOR "FIELD ENGINEERING."

NEW YORK:

JOHN WILEY & SONS.

1882.

PREFACE.

THE object of this work is to reduce the well-known theory of the cubic parabola or multiform compound curve, used as a transition curve, to a practical and convenient form for ordinary field work.

The applicability of this curve to the purpose intended has been fully demonstrated in theory and practice by others, but the method of locating the curve on the ground has been left too much in the mazes of algebra, or else has been described as a system of offsets, or *fudging*. Where a system of deflection angles has been given, the range of spirals furnished has been much too limited for general practice. In consequence the great majority of engineers have contented themselves with locating circular curves only, leaving to the trackman the task of adjusting the track, not to the centres given near the tangent points, but to such an approximation to the spiral as he could give "*by eye*."

The method here described is that of transit and chain, analogous to the method of running circular curves ; it is quite as simple in practice, and as accurate in result. No offsets need be measured, and the curve thus staked out is willingly followed by the trackmen because it "looks right," and is right.

The preliminary labor of selecting a proper spiral for a given case, and of calculating the necessary distances to locate it at the proper place on the line, is here explained, and reduced to the simplest method. Many of

the quantities required have been worked out and tabu-
lated once for all, leaving only those values to be found
which are peculiar to the individual case in hand. A
large number of spirals are thus prepared, and their
essential parts are given in Table III.

In section 22 is developed the method of applying
spirals to existing circular curves, without altering the
length of line, or throwing the track off of the road bed,
an important item to roads already completed. Table
V. contains samples of this kind of work arranged in
order, so that, by a simple interpolation, the proper selection can be made in a given case.

The series of spirals given in Table III. are obtained
by a simple variation of the chord-length, while the deflections and central angles remain constant. This is
the converse of our series of circular curves, in which
the chord is constantly 100 feet, while the deflections
and central angles take a series of values.

The multiform compound curve has been chosen as
the basis of the system, rather than the cubic parabola,
because, while there is no practical difference in the
two, the former is more in keeping with ordinary field
methods, and is far more convenient for the calculation
and tabulation of values *in terms of the chord-unit*, or of
measurement along the curve. While the several com-
ponent arcs of the spiral are thus assumed to be circular, yet the chord-points are points of a true spiral, to
which the track naturally conforms when laid according
to the chord-points given as centres.

The "Railroad Spiral" is in the nature of a sequel to
"Field Engineering;" the same system of notation is
adopted, and any tables referred to, but not given here,
will be found in that work.

WM. H. SEARLES, C. E.

NEW YORK, *July* 1, 1882.

CONTENTS.

CHAPTER I.

INTRODUCTION.

CHAPTER II.

THEORY OF THE SPIRAL.

CHAPTER III.

ELEMENTARY PROBLEMS.

TABLES.

THE RAILROAD SPIRAL.

CHAPTER I.

INTRODUCTION.

1. ON a straight line a railway track should be level
transversely ; on a curve the outer rail should be raised
an amount proportional to the degree of curve. At the
tangent point of a circular curve both of these conditions cannot be realized, and some compromise is usually
adopted, by which the rail is gradually elevated for
some distance on the tangent, so as to gain at the tangent point either the full elevation required for the
curve, or else three-quarters or a half of it, as the case
may be. The consequence of this, and of the abrupt
change of direction at the point of curve, is to give the
car a sudden shock and unsteadiness of motion, as it
passes from the tangent to the curve.

The railroad spiral obviates these difficulties entirely,
since it not only blends insensibly with the tangent on
the one side, and with the circle on the other, but also
affords sufficient space between the two for the proper
elevation of the outer rail. Moreover, since the curvature of the spiral increases regularly from the tangent
to the circle, and the elevation of the outer rail does
the same, the one is everywhere exactly proportional to
the other, as it should be. The use of the spiral allows

the track to remain level transversely for the whole length of the tangent, and yet to be fully inclined for the whole length of the circle, since the entire change in inclination takes place on the spiral.

2. *The office of the spiral* is not to supersede the circular curve, but to afford an easy and gradual transition from tangent to curve, or *vice versa*, in regard both to alignment and to the elevation of the outer rail. A spiral should not be so short as to cause too abrupt a rise in the outer rail, nor yet so long as to render the rise almost imperceptible, and therefore difficult of actual adjustment. Within these limits a spiral may be of any length suited to the requirements of the curve or the conditions of the locality. To suit every case in practice an extensive list of spirals is required from which to select.

CHAPTER II.

THEORY OF THE SPIRAL.

3. THE Railroad Spiral is a compound curve closely resembling the cubic parabola; it is very flat near the tangent, but rapidly gains any desired degree of curvature.

The spiral is constructed upon a series of chords of equal length, and the curve is compounded at the end of each chord. The chords subtend circular arcs, and the degree of curve of the first arc is made the common difference for the degrees of curve of the succeeding arcs. Thus, if the degree of curve of the first arc be o° 10′, that of the second will be o° 20′, of the third, o° 30′, &c.

The spiral is assumed to leave the tangent at the beginning of the first chord, at a tangent point known as the *Point of Spiral*, and designated by the initials *P. S.*, or on the diagrams by the letter S.

4. To determine the co-ordinates of the several chord extremities, let the point S be taken as the origin of co-ordinates, the tangent through S as the axis of Y, and a perpendicular through S as the axis of X. Then *x*, *y*, will represent the co-ordinates of any point of compound curvature in the spiral, *x* being the perpendicular offset from the point to the tangent, and *y* the distance on the tangent from the origin to that offset.

For the purpose of calculation let us assume 100 feet as the chord-length, and o° 10′ as the degree of curve of the first arc of a given spiral. Then, since the degree of curve is an angle at the centre of a circle subtended by a chord of 100 feet, the central angle of the first chord is 10′, of the second 20′, of the third 30′, &c., and the angles which the chords make with the tangent are :

For 1st chord, $\frac{1}{2}$ × 10′ = 5′
" 2d " 10′ + $\frac{1}{2}$ × 20′ = 20′
" 3d " 10′ + 20′ + $\frac{1}{2}$ × 30′ = 45′
" 4th " 10′ + 20′ + 30 + $\frac{1}{2}$ × 40 = 80′
&c., &c., &c.,

or in general the inclination of any chord to the tangent at S is equal to half the central angle subtended by that chord added to the central angles of all the preceding chords. If now we consider the tangent as a meridian, the *latitude* of a chord will be the product of the chord by the cosine of its inclination, and its *departure* will be the product of the chord by the sine of its inclination to the tangent. A summation of the several latitudes for a series of chords will give us the required values of y, and a summation of the several departures will give us the required values of x. By the aid of a table of sines and cosines, we may therefore readily prepare the following statement :

Chord.	Inclin. to tang.	Dep. = 100 sine.	x.	Lat. = 100 cosine.	y.
1	0° 05′	0.145	.145	100.000	100.000
2	0° 20′	0.582	.727	99.998	199.998
3	0° 45′	1.309	2.036	99.991	299.989
4	1° 20′	2.327	4.363	99.979	399.968
&c.			&c.		&c.

In this manner Table I. has been constructed.

5. To calculate the deflection angles of the Spiral; Inst. at S. If in the diagram, Fig. 1, we draw the long chords S2, S3, S4, &c., we may easily determine the angle i, which any long chord makes with the tangent by means of the co-ordinates of the further extremity of the chord, for

$$\tan i = \frac{x}{y}.$$

Having calculated a series of values of the angle i, we may lay out the spiral on the ground by transit deflections from the tangent, the transit being at the point S.

The statement of the calculation is as follows :

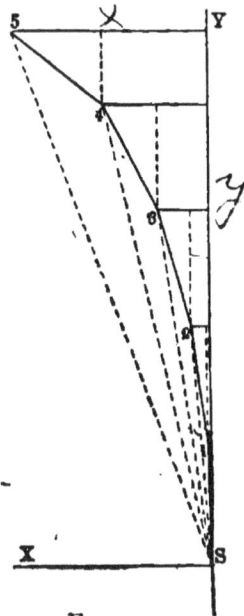

FIG. 1.

Point.	x	y	$\tan i = \frac{x}{y}$.	i
1	.145	100.000	.00145	0° 05′ 00″
2	.727	199.998	.00364	12′ 30″
3	2.036	299.989	.00679	23′ 20″
4	4.363	399.968	.01091	37′ 30″
&c.				&c.

The values of i are more readily found by logarithms however, since

$$\log \tan i = \log x - \log y.$$

By this formula the first part of Table II. (Inst. at S)

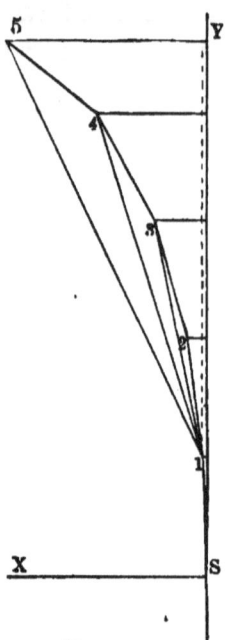

FIG. 2.

has been calculated, and these are the only deflections needed for field use when the entire spiral is visible from S.

6. **To calculate the deflection angles** *when the transit is* **at any other chord-point than S** : Suppose the transit **at point 1**, Fig. 2.

In the diagram draw through the point 1 a line parallel to the tangent at S, and also the long chords 1–3, 1–4, &c., and let a_1 represent the angle between any one of these long chords and the parallel. Then, from the right-angled triangles of the diagram we have the following expressions :

For point 2, $\tan a_1 = \dfrac{x_2 - x_1}{y_2 - y_1} = \dfrac{.572}{99.998} = .00582.$

" " 3, $\tan a_1 = \dfrac{x_3 - x_1}{y_3 - y_1} = \dfrac{1.891}{199.989} = .00945.$

" " 4, $\tan a_1 = \dfrac{x_4 - x_1}{y_4 - y_1} = \dfrac{4.218}{299.968} = .01411.$

&c., &c., &c.

But these are better worked by logarithms, and the values of a_1 found directly from the logarithmic tangent.

Let $s =$ the spiral angle $=$ the angle subtended by any number of spiral chords, beginning at S. Then $s =$ the *sum* of the central angles of the several chords considered ; and it therefore equals the angle between the tangent at S and a tangent at the last point considered. The series of values of the angle s is as follows:

Point.	Angle under single chord.	Angle s.
S	$0°\ 00'$	$0'$
1	$10'$	$10'$
2	$20'$	$30'$
3	$30'$	$1°\ 00'$
4	$40'$	$1°\ 40'$
&c.,		&c.

Since the values of a_1 found above are deflections at point 1 from a parallel to the main tangent, it is evident that if we subtract from each the value of s for point 1, or $10'$, we shall have the deflections, i, from an auxiliary tangent through the point 1, which we require for use in the field. The statement is as follows:

Instrument at point 1 ; $(s = 10')$.

Point.	Angle a_1.	Angle i.
2	$20'$	$10'$
3	$32'\ 30''$	$22'\ 30''$
4	$48'\ 20''$	$38'\ 20''$
&c.,	&c.,	&c.

The instrument will read *zero* on the auxiliary tangent through point 1 where it stands, and of course the back deflection over the circular arc S1 is $05'$. Hence we have the complete table of deflections when the instrument is at point 1.

Similarly, if we suppose the instrument to be **at point 2,** we shall have the statement:

Point.				
3	$\tan a_2 =$	$\dfrac{x_3 - x_2}{y_3 - y_2} =$	$\dfrac{1.309}{99.991} =$	$.01018.$
4	$\tan a_3 =$	$\dfrac{x_4 - x_2}{y_4 - y_2} =$	$\dfrac{3.636}{199.970} =$	$.01527.$
	&c.,			&c.,

and since for point 2, $s = 20'$, we have :

Point.	Angle a_2.	Angle i.
3	$0°\ 35'$	$0°\ 15'$
4	$0°\ 52'\ 30''$	$0°\ 32'\ 30''$
	&c.,	&c.

The instrument will read *zero* on the auxiliary tangent through the point 2, the back deflection to the point 1 is half the central angle under the second chord, or $10'$, and the back deflection to S is the difference between s_2 and the deflection at S for point 2, or $30' - 12'\ 30'' = 17'\ 30''$. We thus may complete the table of deflections for the instrument at point 2.

By a similar process the deflections required at any other chord-point may be deduced. It should be noted, however, in forming the table, that the back deflection to any point is equal to the value of s for the place of the instrument, *less* the value of s for the back-point, *less* the forward deflection at the back-point for the place of the instrument. This is obvious from an inspection of the triangle formed by the two auxiliary tangents and the chord joining the two points in question.

Thus, Fig. 3, when the instrument is at point 4, the back deflection for point 2 is equal to $100' - 30' - 32'\ 30'' = 37'\ 30''.$

In the manner above described has been calculated the complete table of deflections from auxiliary tangents at chord-points, for every chord-point of the spiral up to point 20, Table II. It is evident, that by

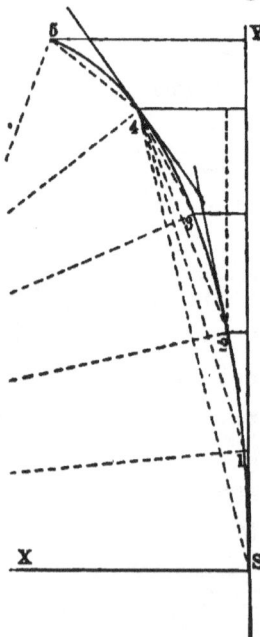

FIG. 3.

means of this table the entire spiral may be located, the transit being set over any chord-point desired, while the chain is carried around the curve in the usual manner; also, that the curve may be laid out in the reverse direction from any chord-point not above the 20th, since all the back deflections are also given.

7. Variation in the chord-length.

We have thus far assumed the spiral to be constructed upon chords of 100 feet, but it is evident that such a spiral would be entirely too long for practical use; it would be 1700 feet long before reaching a $3°$ curve.

We must, therefore, assume a *shorter chord;* but in so doing it will not be necessary to recalculate the *angles and deflections, for these remain the same whatever be the chord-length.* By shortening the chord-length we merely construct the spiral on a smaller scale. The values of x and y and of the radii of the arcs at corresponding points are proportional to the chord-lengths, and the degrees of curve for corresponding chords are (nearly) inversely proportional to the same.

Thus for any chord-length c we have:

$$x : x_{100} :: c : 100, \quad \text{or} \quad x = \frac{c}{100} x_{100}.$$

$$y : y_{100} :: c : 100, \quad \text{or} \quad y = \frac{c}{100} y_{100}.$$

$$R_{\prime} : R_{100} :: c : 100, \quad \text{or} \quad R_{\prime} = \frac{c}{100} R_{100}.$$

Let $D_{\prime} =$ the degree of curve due to radius $R_{\prime}$, and $D_{100} =$ the degree of curve due to radius R_{100}; then,

$$R_{\prime} = \frac{100}{2 \sin \frac{1}{2} D_{\prime}}, \text{ and } R_{100} = \frac{100}{2 \sin \frac{1}{2} D_{100}};$$

whence

$$\sin \tfrac{1}{2} D_{\prime} = \frac{100}{c} \sin \tfrac{1}{2} D_{100},$$

1*

in which $D_{\prime}$ is the degree of curve upon any chord in a spiral of chord-length c, and D_{100} is the degree of curve upon the corresponding chord in the spiral of chord-length 100.

Accordingly, if we assume a chord-length of 10 feet the values of x and y will be $\dfrac{10}{100}$ of those calculated for a chord-length of 100 feet, while the degree of curve on each chord will be (nearly) 10 times as great as before.

8. In the **construction of Table III.**, we have assumed the chord to have every length successively from 10 feet to 50 feet, varying by a single foot, and have calculated the corresponding values of x, y and $D_{\prime}$. The logarithm of x is also added, and the length of spiral nc.

We are thus furnished with 41 distinct spirals, but since the same spiral may be taken with a different number of chords (not less than three) to suit different cases, the variations which the tables furnish amount to no less than 500 spirals, some one or more of which will be adapted to any case that can arise. The maximum length of spiral has been taken at 400 feet; the shortest spiral given is 3×10 feet = 30 feet. Between these limits may be found spirals of various lengths.

9. **The elements of a spiral** are :

D_{n}, The degree of curve on the last chord,

n, The number of chords used,

c, The chord-length,

$n \times c$, The length of spiral,

s, The central angle of the spiral,

x, y, The coordinates of the terminal point.

Every spiral must terminate, or join the circular curve

at a regular chord-point of which the coordinates are known.

10. To select a spiral.

The terminal chord of a spiral must subtend a degree of curve less than that of the circular curve which follows, but the next chord beyond (were the spiral produced) must subtend a degree of curve equal to or differing but a little from that of the circular curve.

Thus, if the circle were a 10 degree curve, the spiral may consist of 5 chords 10 feet long (the degree of curve on the 6th chord being $10° \ 00' \ 45''$), or of 15 chords 26 feet long (the degree of curve on the 16th chord being $10° \ 16' \ 09''$), the length of spiral is 50 feet in one case and 390 in the other; between these limits the tables furnish 15 other spirals of intermediate length, all adapted to join a 10 degree curve.

We may therefore introduce one more condition which will fix definitely the proper spiral to employ. If the *length of spiral* be assumed, we seek in the tables those values of n and c which are consistent with the required value of D_i for $(n + 1)$, at the same time that their *product*, nc, equals as nearly as may be the assumed length of spiral. Thus, if with a 10 degree curve a length of about 130 feet were desirable, we should select either

$$n = 8, \ c = 15, \ D_i = 10° \ 00' \ 45''; \quad nc = 120 \text{ ft.};$$
$$\text{or } n = 9, \ c = 16, \ D_i = 10° \ 25' \ 51''; \quad nc = 144 \text{ ft.}$$

D_i is always taken for $(n + 1)$. When circumstances permit, a chord-length of about 30 feet will give the best proportioned spirals. With a 30 foot chord-length the length of spiral will be about 770 times the superelevation of the outer rail at a velocity of 35 miles per hour.

The value of s depends on the number of chords (n) and is independent of the chord-length. If the angle s were selected from the table, this would fix the number n, and we must then choose the chord-length c so as to give the proper value of D_s. Thus, if s were assumed $= 9° 10'$ then $n = 10$, and $c = 18$ ft. or 19 ft., giving $D_s = 10° 11' 54''$ or $9° 39' 36''$ to suit a 10 degree curve, and making the length (nc) of the spiral either 170 or 180 ft., according to the spiral selected.

The coordinates (x, y) depend on the values of both n and c. They are used in solving the problems of the spiral, being taken directly from Table III. for this purpose, under the value of c and opposite the value of n.

CHAPTER III.

ELEMENTARY PROBLEMS.

11. To find the length C of any long chord beginning at the point of spiral S. Fig. 4. Let L be the other extremity of the long chord, x, y the coordinates of L, and i the deflection angle YSL at S for the point L.

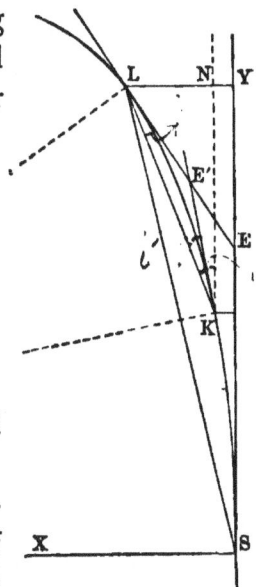

Then
$$C = \frac{y}{\cos i},$$
$$\left.\begin{array}{c} \\ \\ \end{array}\right\} \quad . \quad . \quad . \quad (1.)$$
or
$$C = \frac{x}{\sin i}.$$

The values of x, y and i are found in Tables III. and II.

Example. In the spiral of chord-length = 30 ft. what is the length of the long chord from S to the 10th point?

FIG. 4.

From Table III.,   log x 1.224491
"   " i 3° 12′ 28″ log sin 8.747853

∴ C 299.66 *Ans.*   2.476638

12. To find the lengths of the tangents from the points S and L to their intersection E. Fig. 4. Let x, y be the coordinates of L, and s the spiral angle for the point L.   Then $s =$ the deflection angle between the tangents at E, and

$$LE = \frac{x}{\sin s} \qquad SE = y - x \cot s \quad . \quad . \quad . \quad (2.)$$

The values of x, y and s are found in Tables III. and IV.

Example. In the spiral of chord-length 40 extending to the 9th point, what are the tangents LE and SE ?

From Table III.,		log x	1.219075
" " IV., s 7° 30'		log sin	9.115698
∴ LE $= 126.87$			2.103377
		log x	1.219075
	s 7° 30'	log cot	0.880571
	125.790		2.099646
	y 359.352		
∴ SE $= 233.562$			

13. To find the length C of any long chord KL. Fig. 4. Let x, y be the coordinates of L, and x', y' the coordinates of K ; and let a be the angle LKN which LK makes with the main tangent, and i the deflection angle KLE', and i' the deflection angle LKE'. Then $a = (s - i)$ at the point L, $= (s' + i')$ at K.

$$KL = \frac{KN}{\cos LKN} \qquad \text{or}$$

$$C = \frac{y - y'}{\cos a} \quad . \quad . \quad . \quad . \quad (3.)$$

Example. In the spiral of chord-length 18 what is the

length of the long chord from point 12 to point 20?
Here K = 12 and L = 20 = n.

From Table III., y 346.476

   y' 214.847

_____

   131.629   log   2.119352

From Table II., s' 13°

   i' 10°·07′ 23″

_____

∴ a 23° 07′ 23″ log cos 9.963629

_____

∴ $C = 143.13$ 2.155723

14. To find the lengths of the tangents from any two points L and K to their intersection at E'. Fig. 4. Let s, s' be the spiral angles for the points L and K respectively. Then $(s - s')$ = the deflection angle between tangents at E'. Having first found $C =$ LK by the last problem we have in the triangle LKE'

$$LE' = \frac{C \sin i'}{\sin (s - s')} \qquad KE' = \frac{C \sin i}{\sin (s - s')} \cdot \cdot (4.)$$

Example. In the spiral of chord-length 18 what are the tangents for the points 12 and 20?

By last example, C log   2.155723
From Table IV.,

   $(s - s')$ 35° − 13° = 22° log sin 9.573575

_____

                                2.582148

From Table II., i' 10° 07′ 23″ log sin 9.244927

_____

∴ LE' = 67.15   1.827075

_____

Again :   2.582148
Table II., i 11° 52′ 37″ log sin 9.313468

_____

∴ KE' = 78.635   1.895616

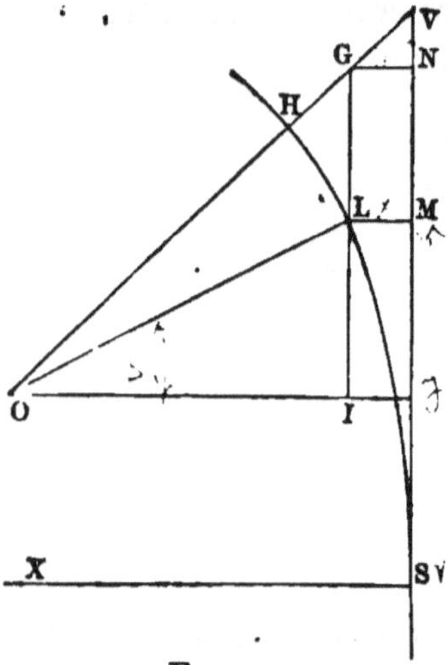

15. Given: *A circular curve and spirals joining two tangents,* to find the tangent distance $T_t = $ VS. Fig. 5.

Let S be the point of spiral, V the intersection of the tangents, SL the spiral, LH one half the circular curve, and O its centre. In the diagram draw GLI parallel to the tangent VS, and GN, LM, and OI perpendicular to VS. Join OL and OV.

FIG. 5.

Then

$$IOL = s; \quad IOV = \tfrac{1}{2}\triangle; \quad OL = R'; \quad SM = y; \quad LM = x.$$

Now
$$SV = SM + NV + MN.$$

But
$$NV = GN \cdot \tan VGN = x \tan \tfrac{1}{2}\triangle.$$

$$MN = GL = OL\frac{\sin LOG}{\sin OGI} = R'\frac{\sin (\tfrac{1}{2}\triangle - s)}{\cos \tfrac{1}{2}\triangle}.$$

Hence
$$T_t = y + x \tan \tfrac{1}{2}\triangle + R'\frac{\sin (\tfrac{1}{2}\triangle - s)}{\cos \tfrac{1}{2}\triangle} \quad . \; . \, (5.)$$

Example. Let the degree of the circular curve be $D' = 7° \, 20'$, and the angle between tangents, $\triangle = 42°$. Let the spiral values be $c = 23$; $n = 9$ $\; . \; . \;$ $s = 7°.30'$. Then by the last equation and the tables,

y		206.627		
x			log	0.978743
$\tfrac{1}{2}\triangle$	21°		log tan	9.584177
		36.55		0.562920

R'	$7°\ 20'$ C		log	2.893118
$\frac{1}{2}\Delta - s$	$13°\ 30'$		log sin	9.368185
$\frac{1}{2}\Delta$	$21°$		a. c. log cos	0.029848

$$ 195.502 2.297151$$

$$\therefore T_{,} = 405.784$$

16. When an **approximate value of** $T_{,}$ is only required we may employ a more convenient formula derived from the fact that the line OI produced bisects the spiral SL very nearly, and that the ordinate to the spiral on the line OI, being only about $\frac{1}{8} x$, may be neglected. Thus,

Approx. $ T_{,} = R' \tan \frac{1}{2}\Delta + \frac{1}{2} nc. . . (6.)$

Example. Same as above.

R'	$7°\ 20'$ C	log	2.893118
$\frac{1}{2}\Delta$	$21°$	log tan	9.584177

$$ 300.1. 2.477295$$

$\frac{1}{2} n c = \frac{1}{2} \times 9 \times 23 103.5$

$$\therefore T_{,} = \text{approx.} 403.6$$

Remark. This formula, eq. (6) when R' is taken equal to the radius corresponding to the degree of curve D, for $(n + 1)$, gives practically correct results. But as in practice, the value of R' will differ somewhat from the radius of $D_{,,}$ so the value of $T_{,}$ derived from this formula will differ more or less from the true value, as in the last example.

17. Given: *the tangent distance* $T_{,} = $ SV, *and the angle* Δ, *and the length of spiral* SL, **to find the radius** R' **of the circular curve,** LH, Fig. 5. The length of spiral is expressed by nc, hence we have from the last equation.

approx., $$R' = (T_t - \tfrac{1}{2}nc) \cot \tfrac{1}{2}\Delta. \quad . \quad . \quad . \quad (7.)$$

After R' is thus found, the values of n and c are to be determined, such that, while their product equals the given length of spiral as nearly as may be, the value of D_t for $(n + 1)$ shall correspond nearly with R'. The values of n and c are quickly found by reference to Table III.

Example. Let $T_t = 406$, $\Delta = 42°$, and $nc = 170$.

$T_t - \tfrac{1}{2}nc$		321		log 2.5065
$\tfrac{1}{2}\Delta$	$21°$			log cot. 0.4158
$\therefore \quad R' =$ say, $6° 51'$ curve,				2.9223

By reference to Table III., we find that when $n = 8$ and $c = 22$, the product nc being 176, the value of D_t for $(n + 1)$ is $6° 49' 19''$, and this is the best spiral to use in this case. But as this spiral is longer than our assumed one, we should decrease the value of R' somewhat, if we would nearly preserve the given value of T_t. For instance, assume $R' =$ radius of $6° 54'$ curve, and using the same spiral, calculate by eq. (4) the resulting value of T_n, and we shall find $T_t = 408.646$.

As this is an exact value of T_t for the values of R', n and c last assumed, and is also a close approximation to the value first given, it will probably answer the purpose completely. If, however, for any reason the precise value of $T_t = 406$ is required, we may find the precise radius which will give it by the following problem.

| **18. Given:** *a curve, and spiral, and tangent-distance,*

$T_{\prime\prime}$ to find the difference in R' corresponding to any small difference in the value of $T_{\prime}$.

If in eq. (5) we assume a *constant spiral*, and give to R' two values in succession and subtract one resulting value of $T_{\prime}$ from the other, we shall find for their difference,

$$\text{diff. } T_{\prime} = \frac{\sin\left(\frac{1}{2}\triangle - s\right)}{\cos\frac{1}{2}\triangle} \text{ diff. } R'. \quad . \quad . \text{ (8.)}$$

Hence

$$\text{diff. } R' = \frac{\cos\frac{1}{2}\triangle}{\sin\left(\frac{1}{2}\triangle - s\right)} \text{ diff. } T_{\prime}. \quad . \quad . \text{ (9.)}$$

Example. When $R' = \text{rad.}$ 6° 54′ curve, $n = 8$, $c = 22$, $T_{\prime} = 408.646$; what radius will make $T_{\prime} = 406$ with the same spiral?

Eq. (9)	diff. $T_{\prime} = 2.646$	log 0.422590
	$\frac{1}{2}\triangle$, 21°	log cos 9.970152
	$(\frac{1}{2}\triangle - s)$, 15°	a. c. log sin 0.587004

∴ diff. R' 9.544   0.979746

R' 6° 54′   830.876

∴ Required radius = 821.332, or 6° 58′ 49″ curve.

Remark. Care must be taken to observe whether in thus changing the value of R', the value of D', the degree of curve, is so far changed as to require a different spiral according to the rule for the selection of spiral, § 10. Should this be the case (which is not very likely), we may adopt the new spiral, and proceed with a new calculation as before.

√ **19. Given** : *a circular curve with spirals joining two tangents,* **to find the external distance** $E_{\prime} = \text{VH}$, Fig. 5.

Let SL be the spiral, LH one-half the circular curve, and O its centre.

Then $VH = VG + GO - OH.$

But $VG = \dfrac{GN}{\cos VGN} = \dfrac{x}{\cos \frac{1}{2}\triangle}$, and in the triangle

GOL, $GO = LO \dfrac{\sin OLI}{\sin LGO} = R' \dfrac{\cos s}{\cos \frac{1}{2}\triangle}$;

$$\therefore \quad E_{\text{\tiny ,}} = \frac{x}{\cos \frac{1}{2}\triangle} + R' \frac{\cos s}{\cos \frac{1}{2}\triangle} - R', \quad . \quad . \quad (\text{10.})$$

or for computation without logarithms

$$E_{\text{\tiny ,}} = \frac{x + R'\,(\cos s - \cos \frac{1}{2}\triangle)}{\cos \frac{1}{2}\triangle}. \quad . \quad . \quad (\text{11.})$$

Example. Let $D' = 7° 20'$, $\triangle = 42°$, and for the spiral let $n = 9$, $c = 23$, giving $s = 7° 30'$, and for $(n + 1)$, $D_s = 7° 15' 04''$.

Eq. (10) x log   0.978743

      $\frac{1}{2}\triangle$ $21°$ a. c. log cos 0.029848

                     10.200   1.008591

      R' $7° 20'$ log   2.893118

      s $7° 30'$ log cos 9.996269

      $\frac{1}{2}\triangle\, 21°$ a. c. log cos 0.029848

                  830.300   2.919235

           sum   840.500

      R' $7° 20'$ 781.840

$\therefore$ $E_{\text{\tiny ,}}$ 58.660

20. Given : *The angle △ at the vertex and the distance* VH = E_n, to determine the radius R' of a circular curve with spirals *connecting the tangents and* passing through the point H. Fig. 5.

Solving eq. (11) for R' we have

$$R' = \frac{E_t \cos \frac{1}{2}\triangle - x}{\cos s - \cos \frac{1}{2}\triangle} \quad \ldots \ldots \quad (12.)$$

But as this expression involves x and s of a spiral dependent on the value of R' we must first find R' approximately, then select the spiral, and finally determine the exact value of R' by eq. (12). The radius R of a simple curve passing through the point H is a good approximation to R'. It is found by eq. (27) Field Engineering:

$$R = \frac{E}{\text{exsec } \frac{1}{2}\triangle},$$

or the degree of curve D may be found by dividing the external distance of a 1° curve for the angle △ by the given value of E_t. But evidently the value of D' will be greater than D, and we may *assume* D' to be from 10′ to 1° greater according to the given value of △, the difference being more as △ is less. We now select from Table III. a value of D_s suited to D' so assumed, and corresponding at the same time to any desired length of spiral. Since D_t so selected corresponds to $(n + 1)$ we take the values of n and x from the next line above D_t in the table, find the value of s from Table IV., and by substituting them in eq. (12) derive the true value of R' for the spiral selected.

Example. Let △ = 42° and E_t = 70, to find the value of R' with suitable spirals.

From table of externals for 1° curve, when △ = 42°
E = 407.64, which divided by 70 gives 5°.823 ; or D =

$5°$ $50'$. Assume D' say $20'$ greater, giving $D' = 6°$ $10'$ approx. If we desire a spiral about 300 feet long we find, Table III., $n = 10$, $c = 30$, and for $(n + 1)$ $D_{,} = 6°$ $06'$ $49''$. For $n = 10$, $s = 9°$ $10'$.

Eq. (12) $\cos \frac{1}{2}\Delta$, $21°$ .93358

       $E_{,}$ 70

              ——————

              65.35060

     x 16.768

              ——————

              48.5826   log 1.686481

$\cos s$ $9°$ $10'$ .98723

$\cos \frac{1}{2}\Delta$ $21°$ .93358   .05365   log 8.729570

       ——————

$\therefore R' = $ rad. (say) $6°$ $20'$ curve. 905.55   2.956911

Proof. Take the exact radius of a $6°$ $20'$ curve and the above spiral and calculate $E_{,}$ by eq. (10) or (11). We shall obtain $E_{,} = 69.97$. *Again:* if we desire a spiral of 200 feet, we find, Table III., $n = 8$, $c = 25$, and for $(n + 1)$ $D_{,} = 6°$, and by eq. (12) $R' = $ rad. of (say) $6°$ $02'$ curve; and by way of proof we find $E_{,} = 69.96$. *Again:* if we desire a spiral of about 400 feet, we find, Table III., $n = 12$, $c = 33$, $s = 13°$, and for $(n + 1)$ $D_{,} = 6°$ $34'$ $07''$. Hence by eq. (12) $R' = $ rad. of (say) $6°$ $50'$ curve. By way of proof we find eq. (10) $E_{,} = 69.95$.

Remark. It is thus evident that a variety of curves with suitable spirals will satisfy the problem, but D' is increased as the spiral is lengthened—for in the example, with a 200 ft. spiral, $D' = 6°$ $02'$; with a 300 ft. spiral, $D' = 6°$ $20'$; and with a 396 ft. spiral, $D' = 6°$ $50'$. Therefore the length of spiral, as well as the value of Δ, must be considered in first assuming the value of D' as compared with D of a simple curve.

21. In case the value of R', as calculated by eq. (12), should give a value to D' inconsistent with the spiral assumed, we may easily ascertain by consulting the table what spiral will be suitable. Choosing a spiral of the same number of chords, but of a different chord-length c, we may calculate R' (a new value) as before ; or the work may be somewhat abbreviated by the following method :

Given : *a change in the value of x,* eq. (12) *to find the corresponding change in the value of R'*; n being constant.

If the values of $E_{,,}$ $\triangle$, and s remain unchanged, we find, by giving to x any two values, and subtracting one resulting value of R' from the other,

$$\text{diff. } R' = \frac{-\text{ diff } x}{\cos s - \cos \frac{1}{2}\triangle} \quad \cdots \quad (13.)$$

that is, R' increases as x decreases, and the differences bear the ratio of $\dfrac{1}{\cos s - \cos \frac{1}{2}\triangle}$.

Example. Let $\triangle = 42°$, $E_{,} = 70$, and for the spiral let $n = 10$, $c = 30$, $s = 9°$ 10', as in the last example, giving $R' = 905.55$; to find the change in R' due to changing c from 30 to 29.

Eq. (13) for $c = 30$, $x = 16.768$
 for $c = 29$, $x = 16.209$

diff. x	·559	log 9.7474
$\cos s - \cos \frac{1}{2}\triangle$ (as before)	.05365	log 8.7296
∴ diff. R'	10.42	1.0178
old value	905·55	
∴ new R'	915·97	$D' =$ (say) 6° 16',

which agrees well with $D_{\iota} = 6° 19' 29''$ for $(n + 1)$ in the new spiral.

If we prove this result by calculating the value of E_{ι} for these new values by eq. (10) we shall find $E_{\iota} =$ 69.93.

The slight discrepancy between these calculated values of E_{ι} and the original is due solely to assuming the value of D' at an exact minute instead of at a fraction.

CHAPTER IV.

SPECIAL PROBLEMS.

22. Given: *two tangents joined by a simple curve*, **to find a circular arc with spirals** *joining the same tangents*, **that will replace the simple curve** *on the same ground as nearly as may be*, **and preserve the same length of line.** Fig. 6.

To fulfill these conditions it is evident that the new curve must be outside of the old one at the middle point H, since the spirals are inside of the simple curve at its tangent points ; also, the radius of the new curve must be less than that of the old one, otherwise the circle passing outside of H would cut the given tangents.

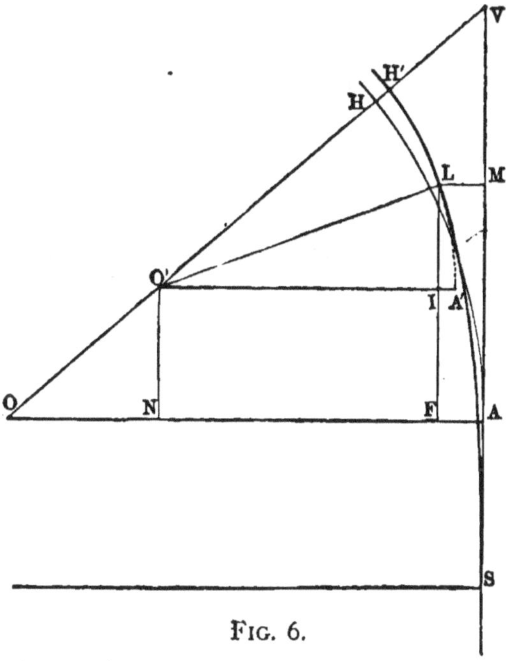

Let SV, Fig. 6 be one tangent, and V the vertex.

FIG. 6.

Let AH be one half the simple curve, and O its centre. Let SL be one spiral, LH' one half the new circular arc, and O' its centre. Draw the bisecting line VO, the radii $AO = R$ and $LO' = R'$, and the perpendicular $LM = x$. Then $MS = y$. Produce the arc H'L to A' to meet the radius $O'A'$ drawn parallel to OA, and let $\frac{1}{2}\Delta$ = the angle AOH = A'O'H'. Let s = the angle A'O'L = the angle of the spiral SL. Let h = the radial offset HH' at the middle point of the curve. Draw O'N and LF perpendicular to OA, LF intersecting $O'A'$ at I.

a. *To find the radius R'* of the new arc LH' in terms of a selected spiral SL.

We have from the figure $AO = ML + FN + NO$. But $AO = R$, $ML = x$, $FN = LO' \cos s = R' \cos s$ and $NO = O'O \cos \frac{1}{2} \Delta = (OH' - O'H') \cos \frac{1}{2} \Delta = (h + R - R') \cos \frac{1}{2} \Delta$; and substituting we have

$$R = x + R' \cos s + (h + R - R') \cos \tfrac{1}{2} \Delta . \quad (14.)$$

whence

$$R' = \frac{R \operatorname{vers} \frac{1}{2}\Delta}{\cos s - \cos \frac{1}{2}\Delta} - \frac{h + \cos \frac{1}{2}\Delta + x}{\cos s - \cos \frac{1}{2}\Delta}. \quad (15.)$$

It is found in practice that h bears a nearly constant ratio to x for all cases under the conditions assumed in this problem. Let k = the ratio $\dfrac{h}{x}$ and the last equation may be written

$$R' = \frac{R \operatorname{vers} \frac{1}{2}\Delta}{\cos s - \cos \frac{1}{2}\Delta} - \frac{(k \cos \frac{1}{2}\Delta + 1)x}{\cos s - \cos \frac{1}{2}\Delta} \quad (16.)$$

which gives the radius of the new arc LH' in terms of s, x and k.

b. *To find the offset $h = \mathrm{HH}'$:*

From eq. (14) we derive

$$h \cos \tfrac{1}{2} \Delta = R\,(1 - \cos \tfrac{1}{2} \Delta) - R'\,(1 - \mathrm{vers}\ s) +$$
$$R' \cos \tfrac{1}{2} \Delta - x$$
$$= R\,(1 - \cos \tfrac{1}{2} \Delta) - R'(1 - \cos \tfrac{1}{2} \Delta) +$$
$$R'\ \mathrm{vers}\ s - x$$
$$= (R - R')\ \mathrm{vers}\ \tfrac{1}{2} \Delta + R'\ \mathrm{vers}\ s - x.$$

Hence

$$h = (R - R')\ \mathrm{exsec}\ \tfrac{1}{2}\ \Delta + \frac{R'\ \mathrm{vers}\ s}{\cos \tfrac{1}{2}\ \Delta} - \frac{x}{\cos \tfrac{1}{2}\ \Delta} \qquad (17.)$$

which gives the value of h in terms of s, x and R'.

c. *To find the value of $d = \mathrm{AS}$:*

We have from the figure $\mathrm{SM} = \mathrm{SA} + \mathrm{NO}' + \mathrm{IL}$. But $\mathrm{SM} = y$, $\mathrm{SA} = d$, $\mathrm{NO}' = \mathrm{OO}' \sin \tfrac{1}{2}\ \Delta$ and $\mathrm{IL} = \mathrm{LO}' \sin s$, and by substitution,

$$y = d + (h + R - R')\ \sin \tfrac{1}{2}\ \Delta + R' \sin s.$$

Hence

$$d = y - [(h + R - R')\ \sin \tfrac{1}{2}\ \Delta + R' \sin s] \quad (18.)$$

which gives the distance on the tangent from the point of curve A to the point of spiral S.

d. *To compare the lengths of the new and old lines :*

$$\mathrm{SAH} = \mathrm{SA} + \mathrm{AH} = d + 100\ \frac{\tfrac{1}{2}\ \Delta}{D}, \quad : \quad . \ (19.)$$

in which D is the degree of curve of AH ;

$$\mathrm{SLH}' = \mathrm{SL} + \mathrm{LH}' = n \cdot c + 100\ \frac{\tfrac{1}{2}\ \Delta - s}{D'} \ (20.)$$

in which D' is the degree of curve of LH'.

If the spiral and arc have been properly selected, the two lines will be of equal length or practically so.

The last two equations assume the circular curves to be measured by 100 foot chords in the usual manner, but when the curves are sharp it is often desirable that they should agree in the *length of actual arcs*, especially where the rail is already laid on the simple curve. For this purpose we use the formulæ

$$\text{SAH (arc)} = d + R \cdot \frac{\Delta}{2} \cdot \frac{\pi}{180} \quad : \qquad . \quad (21.)$$

$$\text{SLH}' \text{ (arc)} = n \cdot c + R' \left(\frac{\Delta}{2} - s \right) \frac{\pi}{180} \quad (22.)$$

in which the angle is expressed in degrees and decimals. If the odd minutes in the angle cannot be expressed by an exact decimal of a degree, the angle should be reduced to minutes, and the divisor of π changed from 180 to 10800.

The value of $\dfrac{\pi}{180}$ is .0174533   log 8.241877

"   " $\dfrac{\pi}{10800}$ is .00029089   " 6.463726.

The length of spiral is given by chord measure in the last equations, since the chords are so short and subtend such small angles that the difference between chord and arc is not material to the problem.

e. *To select a spiral in a given case,* we require to know approximately the value of D', and to select the spiral ($n \cdot c$) such that the value of D, for ($n + 1$) shall not differ greatly from the value of D'. To aid in finding approximate values of D' and k, Table V. has been prepared for curves ranging from $2°$ to $16°$ and central angles ($\triangle$) ranging from $10°$ to $80°$.

Assume s at pleasure (less than $\frac{1}{2}$ $\triangle$), which fixes the value of n. Then inspect Table V. opposite n for values of D and $\triangle$ next above and below the values of D and $\triangle$ in the given problem, and by inference or interpolation decide on the probable values of k and D'. Then in Table III. select that value of c which gives D, for $(n + 1)$ most nearly agreeing with D'. Now calculate R' by eq. (16), and as this will usually give the degree of curve D' fractional, take the value of D' to the *nearest minute* only, and assume the corresponding value of R' as the *real* value of R'. A table of radii makes this operation very simple.

But should it happen that D' differs too widely from from $D_{s(n+1)}$ to make an easy curve, increase or diminish the chord-length c by 1, thus giving a new value to x in eq. (16), and also a new value of $D_{s(n+1)}$ with which to compare the resulting D'. In changing x only the last term of eq. (16) is affected, and the first term does not require recalculation.

f. When the value of R' is decided, substitute it in eq. (17) and calculate h. But if it happens that the value of R' selected differs not materially from the result of eq. (16), we have at once $h = kx$; or in case the value of R' is changed considerably from the result of eq. (16), the corresponding change in h will be

$$\text{diff. } h = -\frac{\cos s - \cos \frac{1}{2}\triangle}{\cos \frac{1}{2}\triangle} \text{ diff. } R', \cdot (22\tfrac{1}{2})$$

which may therefore be applied as a correction to $h = kx$, and we thus avoid the use of eq. (17). Eq. ($22\frac{1}{2}$) is derived from eq. (15) by supposing h to have any two values, and subtracting the resulting values of R' from each other.   Note that h diminishes as R' increases, and *vice versa.*

When R' and h are found, proceed to find d by eq. (18), and the length of lines by eq. (19), (20), or by (21), (22), as may be preferred.   But to produce equality of actual arcs, k must be a little greater than when equality by chord-measure is desired.

Should the lines not agree in length so nearly as desired, a change of one minute $\pm$ in the value of D' may produce the desired result, but any such change necessitates, of course, a recalculation of h and d.

The values of k in Table V. appear to vary irregularly.   This is due to the selection of D' to the nearest minute, and also to the choice of spiral chord-lengths, c, not in an exact series.   The reader is recommended to supplement this table by a record of the problems he solves, so that the values of R' and k may be approximated with greater certainty.

Example.   Given a 6° curve, with a central angle of $\Delta = 50° \, 12'$, to replace it by a circular arc with spirals, preserving the same length of line.   Assume $s = 7° \, 30'$ giving $n = 9$.

Since 6° is an average of 4° and 8°, while 50° 12' is nearly an average of 40° and 60°, we examine Table V. under 4° curve and 8° curve, and opposite $\Delta = 40°$ and 60° on the same line as $s = 7° \, 30'$, and take an average of the four values of $D_{s\,(n+1)}$, thus found ; also of the four values of k ; we thus find *approx.* $k = .0885$, and $D' = 6° \, 18' \pm$.   Now looking in Table III., opposite $n = 9$, we find that when $c = 26$, $D_{s\,(n+1)} = 6° \, 24' \, 48''$, we therefore assume $c = 26$, and proceed to calculate R' by eq. (16).

Eq. (16) $\cos s\, 7° 30'$ ·99144
$\cos \tfrac{1}{2}\triangle$ 25° 06'   ·90557

	.08587	a. c. log 1.066159
R 6°		log 2.980170
vers $\tfrac{1}{2}\triangle$ 25° 06'		log 8.975116
	1050.6	log 3.021445
$\cos s - \cos \tfrac{1}{2}\triangle$ •		a. c. log 1.066159
$1 + k \cos \tfrac{1}{2}\triangle = 1.080$		0.033424
x		1.031989
	135.4	2.131572
$\therefore R'$ (say 6° 16')	915.2	

Eq. (17) R 6°   955.366
       R' 6° 16'   914.750

$(R - R')$	40.616	log 1.608697
exsec $\tfrac{1}{2}\triangle$ 25° 06'		log 9.018194
	4.235	log 0.626891
R' 6° 16'		log 2.961303
vers s , 7° 30'		log 7.932227
$\cos \tfrac{1}{2}\triangle$ 25° 06'		a. c. log 0.043079
	8.642	log 0.936609
	12.877	
x		log 1.031989
$\cos \tfrac{1}{2}\triangle$ 25° 06'		a. c. log 0.043079
	11.887	1.075068
$\therefore h$	0.990	
Eq. (18) $(R - R')$	40.616	
	41.606	log 1.619156
$\sin \tfrac{1}{2}\triangle$ 25° 06'		log 9.627570
	17.649	log 1.246726

R' $6°\ 16'$ log   2.961303

$\sin s\ 7°\ 30'$ log   9.115698

               119.399   log   2.077001

             . 137.048

y 233.579

$\therefore d$ 96.531

Eq. (19) $\dfrac{25.1° \times 100}{6} =$ 418.333

$\therefore$ SAH   514.864

Eq. (20) $(\frac{1}{2}\triangle - s) = 1056' \times 100$ log   5.023664

D' $376'$ log   2.575188

               280.851   log   2.448476

$n.c$ 9×26 234.

$\therefore$ SLH$'$ 514.851

Difference $-$.013

actual $k = \dfrac{h}{x} = 0.092$

Comparison of actual arcs.

Eq. (21) $25.1°$ log 1.399674 | Eq. (22) $17.6°$ log 1.245513

           $1°$ log 8.241877 | $1°$ log 8.241877

R $6°$ log 2.980170 | R' $6°\ 16'$ log 2.961303

    418.525   log 2.621721 | 280.991 log 2.448693

a 96.531 | $n.c$ 234.

    515.056 | 514.991

                       Difference $= -$ 0.065

23. Given : *a simple curve joining two tangents,* **to move the curve inward** *along the bisecting line* VO **so that it may join a given spiral without change of radius.** Fig. 7.

Let SL be the given spiral, AH one-half of the given curve, and HL a portion of the same curve in its new position, and compounded with the spiral at L.

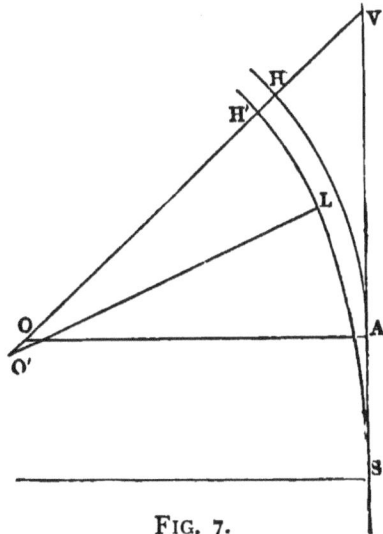

To find the distance $h = \mathrm{HH}' = \mathrm{OO}'$:

Since the new radius is equal to the old one, or $R' = R$, we have from eq. (17) by changing the sign of h, since it is taken in the opposite direction,

FIG. 7.

$$h = \frac{x - R \operatorname{vers} s}{\cos \tfrac{1}{2} \Delta} \quad . \quad . \quad . \quad . \quad . \quad . \quad (23.)$$

To find the distance $d = \mathrm{AS}$:

Changing the sign of h in eq. (18) and making $R' = R$ we have

$$d = y - (R \sin s - h \sin \tfrac{1}{2} \Delta) \quad . \quad . \quad . \quad (24.)$$

This problem is best adapted to curves of large radius and small central angle.

Example. Given, a curve $D = 1° \, 40'$ and $\Delta = 26° \, 40'$, and a spiral $s = 1°$, $n = 3$, and $c = 40$, to find h and d and the length LH'.

Eq. (23) $R \; 1° \, 40'$		log 3.5363
vers s \quad $1°$		log 6.1827
$\cos \tfrac{1}{2} \Delta \; 13° \, 20'$		a. c. log 0.0119

2*

$$\cdot538 \quad \log \quad 9.7309$$

$$x \qquad\qquad\qquad \log \quad 9.9109$$
$$\cos \tfrac{1}{2}\Delta' \qquad\qquad \text{a. c. log} \quad 0.0119$$

$$\cdot837 \qquad\qquad 9.9228$$

$$\therefore h \qquad \cdot299$$

Eq. (24) $\quad R\ 1^\circ\ 40' \qquad\qquad \log \quad 3.536289$
$$\sin\ s\ 1^\circ \qquad\qquad\quad `` \quad 8.241855$$

$$59.999 \qquad\qquad 1.778144$$

$$h \qquad\qquad \cdot299 \qquad \log \quad 9.4757$$
$$\sin \tfrac{1}{2}\Delta\ 13^\circ\ 20' \qquad\qquad\quad `` \quad 9.3629$$

$$\cdot069 \qquad\qquad 8.8386$$

$$59.930$$
$$y' \qquad\qquad 119.996$$

$$\therefore d \qquad\qquad 60.066$$

$$H'O'L = (\tfrac{1}{2}\Delta - s) = 12^\circ\ 20' \quad \therefore \quad H'L = 740 \text{ feet.}$$

24. Given, *a simple curve joining two tangents*, **to compound the curve near each end** *with an arc and spiral joining the tangent* **without disturbing the middle portion of the curve.** Fig. 8.

Let H be the middle point of the given curve, Q the point of compounding with the new arc, and L the point where the new arc joins the spiral SL.

Let $s =$ the spiral angle, and let $0 = $ AOQ. Now in this figure AOQS will be analogous to AOH'S of Fig. 6, if in the latter we suppose H' to coincide with H or $h = 0$. If, therefore, in eq. (15) we write 0 for $\tfrac{1}{2}\Delta$ and make $h = 0$, we have for the new radius O'Q, in terms of 0 and the spiral assumed. But as the value of D' resulting is likely to be fractional and must be adhered to, it is preferable to assume R' a little less than R, select a suitable spiral and calculate the angle 0. Resolving eq. (17) after making $h = o$ and replacing $\frac{1}{2} \triangle$ by 0, we have

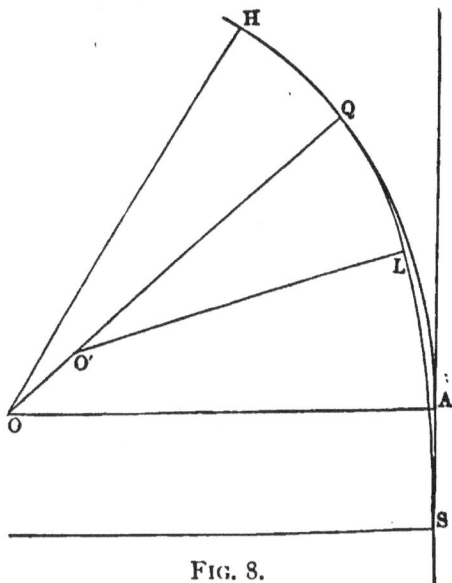

FIG. 8.

$$\text{vers } 0 = \frac{x - R' \text{ vers } s}{R - R'} \quad \ldots \ldots \ldots \quad (26.)$$

The angle 0 so found must be less than $\frac{1}{2} \triangle$, and indeed for good practice should not exceed $\frac{1}{3} \triangle$. If too large, 0 may be reduced by assuming a smaller value of R', and repeating the calculation with a suitable spiral. Otherwise it will be preferable to use one of the foregoing problems in place of this. This problem is specially useful when the central angle is very large.

To find the distance $d = AS$, we have only to write 0 for $\frac{1}{2} \triangle$ and make $h = o$ in eq. (18), whence

$$d = y - [(R - R') \sin 0 + R' \sin s] \quad \ldots \quad (27.)$$

Example. Given a curve $D = 2° \, 30'$, $\triangle = 35°$, to compound it with a curve $D' = 2° \, 40'$ and a spiral $s = 2° \, 30'$, $n = 5$, $c = 37$.

Eq. (26) R　2° 30′　2292.01
　　　　 R'　2° 40′　2148.79

$R - R'$	143.22		log 2.156004
x			log 0.471203
	.020663		log 8.315199

$R - R'$		a. c. log 7.843996
vers s　2° 30′		log 6.978536
R'　　2° 40′		· log 3.332193
	.014280	log 8.154725

∴ vers θ 6° 28′ 30″　　.006383

Eq. (27) $R - R'$　　　　　　log 2.156004
　　sin θ　6° 28′ 30″　　　　9.052192

　　　　　　　　　16.151　　　1.208196

R'　　2° 40′　　　　　　　　3.332193
sin s　2° 30′　　　　　　　　8.639680

　　　　　　　　　93.729　　　1.971873

　　　　　　　　　109.880
y　　　　　　　　184.962

∴ d　　　　　　　75.082
　 AH　　　　　　　700.
　　　　　　　　　————　775.082
SL, $= n . c =$　　　185.00
LQ, $\theta - s =$　3° 58′ 30″　149.06
QH, $\frac{1}{2} \triangle - \theta =$ 11° 01′ 30″ 441.00　775.060

　　　　　Difference　　　— .022

25. **Given : a compound curve joining two tangents, to replace it by another with spirals, preserving the same length of line.** Fig. 9.

Let $\Delta_2 = AO_2P$, the angle of the arc AP, and $\Delta_1 = PO_1B$, the angle of the arc PB. Let $R_2 = AO_2$, and $R_1 = BO_1$.

Adopting the method of § 22, the offset h must be made at the point of compound curve P instead of at the middle point. Considering first the arc of the larger radius AO_2, the formulæ of §22 will be made to apply to this case by writing Δ_2 in place of $\frac{1}{2}\Delta$, and R_2 in place of R, whence eq. (16)

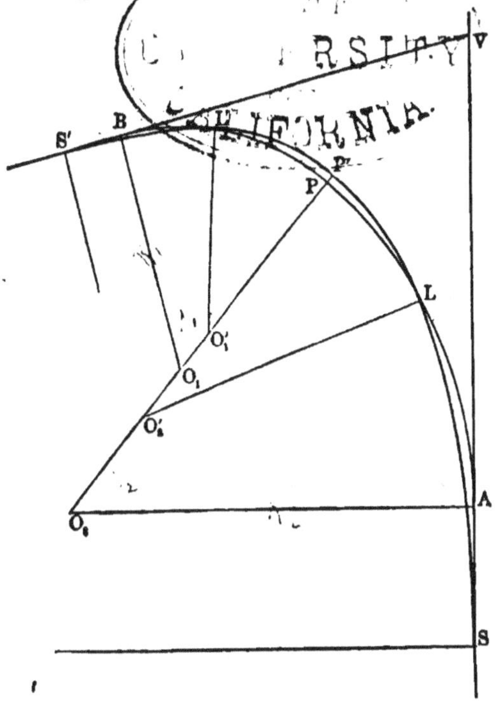

FIG. 9.

$$R_2' = \frac{R_2 \text{ vers } \Delta_2}{\cos s - \cos \Delta_2} - \frac{(k \cos \Delta_2 + 1)\, x}{\cos s - \cos \Delta_2} \quad \ldots \quad (28.)$$

and eq. (17)

$$h = (R_2 - R_2') \text{ exsec } \Delta_2 + \frac{R_2' \text{ vers } s}{\cos \Delta_2} - \frac{x}{\cos \Delta_2} \quad (29.)$$

and eq. (18)

$$d = y - \left[(h + R_2 - R_2') \sin \Delta_2 + R_2' \sin s \right] \quad \ldots \quad (30.)$$

But in considering the second arc PB, we must retain the value of h already found in eq. (29) in order that the arcs may meet in P'. We therefore use eq. (15) which, after the necessary changes in notation, becomes

$$R_1' = \frac{R_1 \text{ vers } \triangle_1}{\cos s - \cos \triangle_1} - \frac{h \cos \triangle_1 + x}{\cos s - \cos \triangle_1}, \quad \ldots \quad (31.)$$

which value of R_1' must be adhered to.

The spiral selected for use in the last equation is independent of the spiral just used in connection with R_2'. It should be so selected that while suitable for R_1' its value of x may be equal to $\frac{h}{k}$ as nearly as may be, the value of k being inferred from Table V. for D' and $2 \triangle_1$.

Assuming the value of R_1' found by eq. (31), even though D_1' be fractional, we may verify the value of h by

$$h = (R_1 - R_1') \text{ exsec } \triangle_1 + \frac{R_1' \text{ vers } s}{\cos \triangle_1} - \frac{x}{\cos \triangle_1} (32.)$$

and then proceed to find $d' = $ BS' by

$$d' = y - [(h + R_1 - R_1') \sin \triangle_1 + R_1' \sin s] \quad (33.)$$

Example. Given the compound curve $D_1 = 8°.$, $\triangle_1 = 29°$ and $D_2 = 6°$, $\triangle_2 = 25°06'$: to replace it by another compound curve connected with the tangents by spirals.

Considering first the 6° branch of the curve, we may assume the spiral $s = 7°30'$, $n = 9$, $c = 26$. This part of the problem is then identical with the example given in § 22, by which we find $h = .990$ and $d = 96.531$.

To select a spiral for the 8° branch, having reference at the same time to this value of h; we find in Table V.

under $D = 8°$ and opposite $\triangle = 2\,\triangle_1 = 58°$ or say 60°, that the given value of h falls between the tabular values of h for $nc = 9 \times 20$, and $nc = 10 \times 22$. We therefore infer that the spiral $nc = 9 \times 21$ is most suitable to this case. Adopting this, we have

Eq. (31) $\cos s\ 7°30'$ .99144
$\cos \triangle_1\ 29°.87462$

		.11682	log 9.067517	a.c. log	0.932483
R_1	8°		"		2.855385
vers $\triangle_1 29°$			"		9.098229

		769.302	"	2.886097
$h \cos\ 29°$	.866			
x	8.694			

	9.560	"	0.980458
$\cos s - \cos \triangle_1$		a.c. "	0.932483

	81.835	" 1.912941

$\therefore R_1'\ 8°20'30'$ 687.467
Eq. (33) $(h + R_1)$ 717.769

	30.302	" 1.481471
$\sin \triangle_1 29°$		" 9.685571

	14.691	" 1.167042

R_1' 687.467   " 2.837251
$\sin s\ 7°30'$ 9.115698

	89.732	1.952949

104.423
188.660

$\therefore\ \dfrac{y}{d}$ 84.237

For the methods of computing the lengths of lines,
see § 22.

26. Given : a compound curve *joining two tangents,*
to move the curve inward *along the line* PO_2 *so that*
spirals may be introduced **without changing the ra-**
dii. Fig. 10.

The distance $h = PP'$ is found for the arc of larger

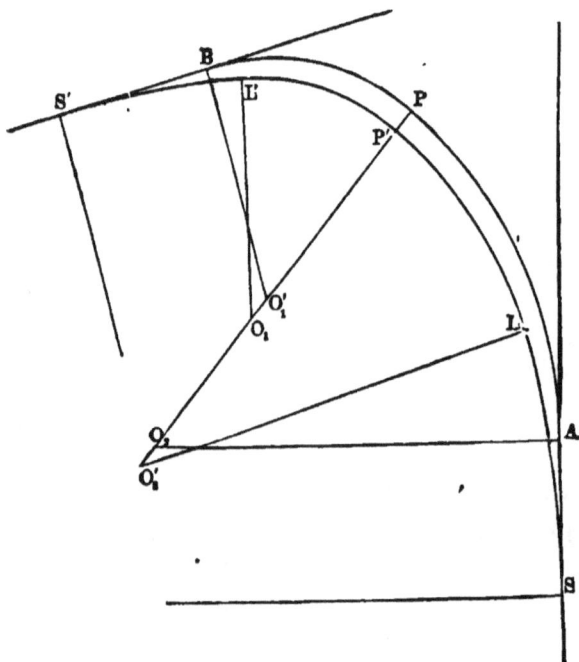

Fig. 10.

radius AO_2 by the following formula derived by analogy
from eq. (23):

$$h = \frac{x - R_2 \operatorname{vers} s}{\cos \Delta_2} ; \quad . \quad . \quad . \quad (34.)$$

and for the distance $d = AS$ we have analogous to eq.
(24):

$$d = y - (R_2 \sin s - h \sin \Delta_2) \quad . \quad (35.)$$

. Now the same value of h, found by eq. (34) must be used for the arc PB, and a spiral must be selected which will produce this value. To find the proper spiral, we have from eq. (34) after changing the subscripts,

$$x = R_1 \text{ vers } s + h \cos \triangle_1 \quad . \quad . \quad (36.)$$

The last term is constant. The values of x and s must be consistent with each other, and approximately so with the value of R_1. Assume s at any probable value, and calculate x by eq. (36). Then in Table III. look for this value of x opposite n corresponding to s, and note the corresponding value of the chord-length c. Compare D_2 of the table with D_1 and if the disagreement is too great select another value of s and proceed as before.

The term R_1 vers s may be readily found, and with sufficient accuracy for this purpose, by dividing the value of R $1°$ vers s Table IV. by D_1. If the calculated value of x is not in the Table III., it may be found by interpolating values of c to the one tenth of a foot, since for a given value of s or n the values of x and y are proportional to the values of c.

When the proper spiral has been found and the value of c determined, it only remains to find the value of $d =$ BS' by

$$d = y - (R_1 \sin s - h \sin \triangle_1), \quad . \quad (37.)$$

in which the value of y will be taken according to the values of c and s just established.

Example. Given: $D_2 = 1°40'$, $\triangle_2 = 13°20'$, $D_1 = 3°$, and $\triangle = 22°40'$, to apply spirals without change of radii. Fig. 10.

Assume for the $1°$ $40'$ arc the spiral $s = 1°$, $n = 3$, $c = 40$. This part of the problem is then identical with the example given in § 23, from which we find $h = 0.299$.

For the second part, if we assume $s = 1° 40'$, $n = 4$, and find by Table IV. R_1 vers $s = \dfrac{2.424}{3} = 0.808$, we have by eq. (36)

$$x = 0.808 + 0.277 = 1.085,$$

the nearest value to which in Table III. is under $c = 25$, giving $D_s = 2° 40'$, or for $(n + 1)$, $D_s = 3° 20'$, which is consistent with $D_1 = 3°$. By interpolation we find that our value of x corresponds exactly to $c = 24.85$, $n = 4$, and therefore the spiral should be laid out on the ground by using this precise chord.

In order to find $d = BS'$ we first find the value of y by interpolation for $c = 24.85$, when by eq. (37) we have

$$d = 99.391 - (55.554 - 0.115) = 43.952.$$

27. Given : a compound curve *joining two tangents,* **to introduce spirals without disturbing the point of compound curvature P. Fig. 11.**

a. *The radius of each arc may be shortened,* giving two new arcs compounded at the same point P. Having selected a suitable spiral, we have for the arc AP by analogy from eq. (15), since $h = 0$,

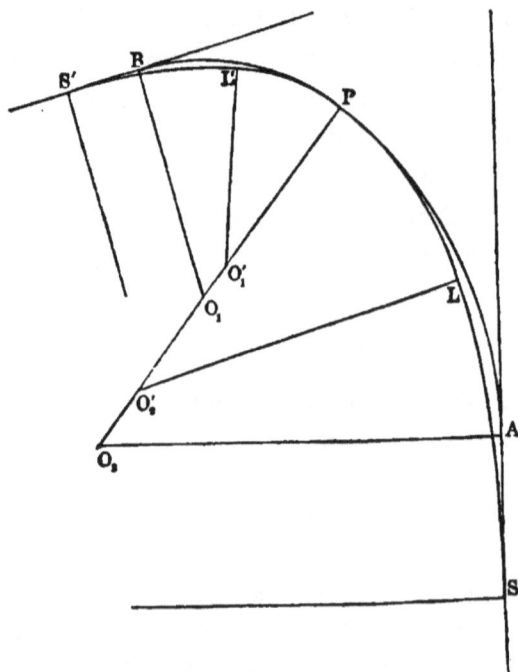

Fig. 11.

$$R_2' = \frac{R_2 \text{ vers } \Delta_2 - x}{\cos s - \cos \Delta_2}; \quad \cdots \cdots \quad (38.)$$

and, similarly, after selecting another spiral for the arc PB,

$$R_1' = \frac{R_1 \text{ vers } \Delta_1 - x}{\cos s - \cos \Delta_1} \quad \cdots \cdots \quad (39.)$$

From eq. (18) we have for the distance AS,

$$d = y - [(R_2 - R_2') \sin \Delta_2 + R_2' \sin s], \quad . \ (40.)$$

and for the distance BS',

$$d = y - [(R_1 - R_1') \sin \Delta_1 + R_1' \sin s] \ . \ (41.)$$

The values of D_1' and D_2' resulting from eq. (39) and (40) must be adhered to, even though involving a fraction of a minute.

b. *Either arc may be again compounded* at some point Q, leaving the portion PQ undisturbed, as explained in § 24. Fig. 12.

Let $0 =$ the angle AO_2Q, and we have from eq. (26), after selecting a suitable spiral and assuming R_2',

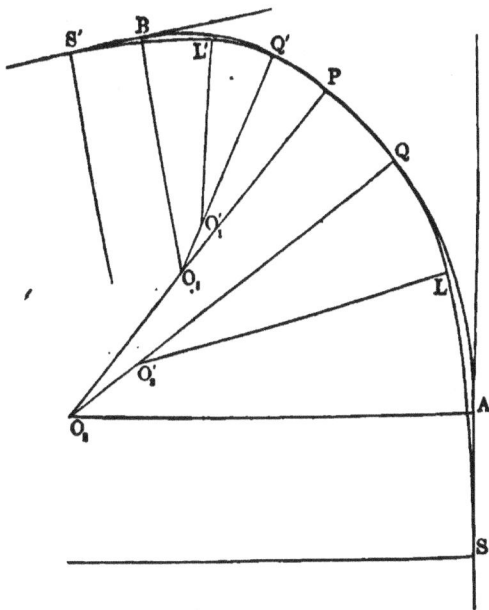

Fig. 12.

$$\text{vers } 0 = \frac{x - R_2' \text{ vers } s}{R_2 - R_2'} \quad \cdots \cdots \quad (42.)$$

For the distance AS, we have from eq. (27)

$$d = y - [(R_2 - R_2') \sin \theta + R_2' \sin s] \quad . \quad (43.)$$

Similar formulæ will determine the angle $\theta = BO_1Q'$ and the distance BS' for the other arc PB in terms of a suitable spiral : thus,

$$\text{vers } \theta = \frac{x - R_1' \text{ vers } s}{R_1 - R_1'} \quad . \quad . \quad . \quad . \quad (44.)$$

$$d = y - [(R_1 - R_1') \sin \theta + R_1' \sin s] \quad . \quad (45.)$$

The method **a** may be adopted with one arc and the method **b** with the other if desired, since the point P is not disturbed in either case. The former is better adapted to short arcs, the latter to long ones.

These methods apply also to compound curves of more than two arcs, only the extreme arcs being altered in such cases.

CHAPTER V.

FIELD WORK.

28. Having prepared the necessary data by any of the preceding formulæ, the engineer· locates the point S on the ground by measuring along the tangent from V or from A. He then places the transit at S, makes the verniers read *zero*, and fixes the cross-hair upon the tangent. He then instructs the chainmen as to the proper chord c to use in locating the spiral, and as they measure this length in successive chords, he makes in succession the deflections given in Table II. under the heading "Inst. at S," lining in a pin or stake at the end of each chord in the same manner as for a circle.

When the point L is reached by (n) chords, the transit is brought forward and placed at L; the verniers are made to read the first deflection given in Table II. under the heading "Inst. at n" (whatever number n may be), and a backsight is taken on the point S. If the verniers are made to read the succeeding deflections, the cross-hair should fall successively on the pins already set, this being merely a check on the work done, until when the verniers read *zero*, the cross-hair will define the tangent to the curve at L. From this tangent the circular arc which succeeds may be located in the usual manner.

In case it became necessary to bring forward the transit before the point L is reached, select for a transit-point the extremity of any chord, as point 4, for example, and setting up the transit at this point, make the verniers read the first deflection under " Inst. at 4," Table II., and take a backsight on the point S.   Then, when the reading is zero, the cross-hair will define the tangent to the curve at the point 4, and by making the deflections which follow in the table opposite 5, 6, &c., those points will be located on the ground until the desired point L is reached by n chords from the beginning S.

The transit is then placed at L, and the verniers set at the deflection found under the heading " Inst. at n " (whatever number n may be), and opposite (4) the point just quitted.   A backsight is then taken on point 4, and the tangent to the curve at L found by bringing the zeros together, when the circular arc may be proceeded with as usual.

29. *To locate a spiral from the point* L *running toward the tangent at* S: we have first to consider the number of chords (n) of which the spiral SL is composed.   Then, placing the transit at L, reading *zero* upon the tangent to the curve at L, look in Table II. under the heading " Inst. at n," and make the deflection given just *above* $0° 00'$ to define the first point on the spiral from L toward S ; the next deflection, *reading up the page*, will give the next point, and so on till the point S is reached.

The transit is then placed at S ; the reading is taken from under the heading "Inst. at S," and on the line n for a backsight on L.   Then the reading zero will give the tangent to the spiral at the point S, which should coincide with the given tangent.

If S is not visible from L, the transit may be set up at any intermediate chord-point, as point 5, for example. The reading for backsight on L is now found under the

heading " Inst. at 5," and on the line *n* corresponding to
L ; while the readings for points between 5 and S are
found *above* the line 5 of the same table. The transit
being placed at S, the reading for backsight on 5, the
point just quitted, is found under " Inst. at S " and
opposite 5, when by bringing the zeros together a tangent to the spiral at S will be defined.

30. Since the spiral is located exclusively by its
chord-points, if it be desired *to establish the regular* 100-
foot stations as they occur upon the spiral, these must be
treated as *plusses* to the chord-points, and a deflection
angle will be interpolated where a station occurs. *To
find the deflection angle for a station succeeding any chord-
point :* the differences given in Table II. are the deflections over one chord-length, or from one point to the
next. For any intermediate station the deflection will
be assumed proportional to the sub-chord, or distance
of the station from the point. We therefore multiply
the tabular difference by the sub-chord, and divide by the
given chord-length, for the deflection from that point to
the station. This applied to the deflection for the point
will give the total deflection for the station.

This method of interpolation really fixes the station
on a circle passing through the two adjacent chord-
points and the place of the transit, but the consequent
error is too small to be noticeable in setting an ordinary
stake. Transit centres will be set only at chord-points,
as already explained.

31. It is important that the spiral should join the
main tangent *perfectly*, in order that the full theoretic
advantage of the spiral may be realized. In view of
this fact, and on account of the slight inaccuracies
inseparable from field work as ordinarily performed, it
is usually preferable to establish carefully the two points

of spiral S and S' on the main tangents, and beginning at each of these in succession, locate the spirals to the points L and L'. The latter points are then connected by means of the proper circular arc or arcs. Any slight inaccuracy will thus be distributed in the body of the curve, and the spirals will be in perfect condition.

32. A spiral may be located without deflection angles, by simply laying off in succession the abscissas y and ordinates x of Table III. corresponding to the given chord-length c. The tangent EL at any point L, Fig. 4, is then found by laying off on the main tangent the distance YE $= x$ cot s, and joining EL. In using this method the chord-length should be measured along the spiral as a check.

33. In making the final location of a railway line through a smooth country the spirals may be introduced at once by the methods explained in Chapter III. But if the ground is difficult and the curves require close adjustment to the contour of the surface, it will be more convenient to make the study of the location in circular curves, and when these are likely to require no further alterations, the spirals may be introduced at leisure by the methods explained in Chapter IV. The spirals should be located before the work is staked out for construction, so that the road-bed and masonry structures may conform to the centre line of the track.

34. When the line has been first located by circular curves and tangents, a description of these will ordinarily suffice for right of way purposes ; but if greater precision is required the description may include the spirals, as in the following example :

" Thence by a tangent N. 10° 15'E., 725 feet to station 1132 + 12; thence curving left by a spiral of 8 chords, 288 feet to station 1135; thence by a 4° 12' curve (radius

1364.5 feet), 666.7 feet to the station 1141 + 66.7; thence by a spiral of 8 chords 288 feet to station 1144 + 54.7 P.T. Total angle 40° left. Thence by a tangent N. 29° 45' W.," &c.

35. When the track is laid, the outer rail should receive a relative elevation at the point L suitable to the circular curve at the assumed maximum velocity. Usually the track should be level transversly at the point S, but in case of very short spirals, which sometimes cannot be avoided, it is well to begin the elevation of the rail just one chord-length back of S on the tangent.

36. Inasmuch as the perfection of the line depends on adjusting the inclination of the track proportionally to the curvature, and in *keeping it so*, it is extremely important that the points S and L of each spiral should be secured by permanent monuments in the centre of the track, and by witness-posts at the side of the road. The posts should be painted and lettered so that they may serve as guides to the trackmen in their subsequent efforts to grade and "line up " the track. The post opposite the point S may receive that initial, and the post at L may be so marked and also should receive the figures indicating the degree of curve.

37. The field notes may be kept in the usual manner for curves, introducing the proper initials at the several points as they occur. The chord-points of the spiral may be designated as *plusses* from the last regular station if preferred, as well as by the numbers 1, 2, 3, &c., from the point S. Observe that the chord numbers always begin at S, even though the spiral be run in the opposite direction.

ELEMENTS OF THE SPIRAL

Point n.	Degree of curve Ds.	Spiral angle s.	Inclina- tion of chord to axis of Y. 	Latitude of each chord. 100 × cos Incl.	Sum of the lati- tudes, y.
0	0° 00'	0° 00'	0° 00'		
I	10'	10'	05'	99.99989423	99.99989423
2	20'	30'	20'	99.99830769	199.99820192
3	30'	1°	45'	99.99143275	299.98963467
4	40'	1° 40'	1° 20'	99.97292412	399.96255879
5	50'	2° 30'	2° 05'	99.93390007	499 89645886
6	1°	3° 30'	3°	99.8629535	599.7594123
7	1° 10'	4° 40'	4° 05'	99.7461539	699.5055662
8	1° 20'	6°	5° 20'	99.5670790	799.0726452
9	1° 30'	7° 30'	6° 45'	99.3068457	898.3794909
10	1° 40'	9° 10'	8° 20'	98.944164	997.3236549
11	1° 50'	11°	10° 05'	98.455415	1095.779070
12	2°	13°	12°	97.814760	1193.593830
13	2° 10'	15° 10'	14° 05'	96.994284	1290.588114
14	2° 20'	17° 30'	16° 20'	95.964184	1386.552298
15	2° 30'	20°	18° 45'	94.693014	1481.245312
16	2° 40'	22° 40'	21° 20'	93.147975	1574.393287
17	2° 50'	25° 30'	24° 05'	91.295292	1665.688579
18	3°	28° 30'	27°	89.100650	1754.789229
19	3° 10'	31° 40'	30° 05'	86.529730	1841.318959
20	3° 20'	35°	33° 20'	83.548780	1924.867739

			Point n.	$\text{Log}\,\dfrac{x}{y} =$ log tan i.	Deflection angle, i.
			I	7.1626964	0° 05' 00."00
			2	7.5606380	0° 12' 30."00
			3	7.8317091	0° 23' 20."00
			4	8.0377730	0° 37' 29."99
			5	8.2041217	0° 54' 59."97
			6	8.3436473	1° 15' 49."90
			7	8.4638309	1° 39' 59."75
			8	8.5694047	2° 07' 29."45
			9	8.6635555	2° 38' 18."90
			10	8.7485340	3° 12' 27."95

Departure of each chord.	Sum of the departures,	Logarithm,	Logarithm,	Point
100 × sin Incl.	*x*.	log *y*.	log *x*.	*n*.
				0
.1454441	.1454441	1.9999995	9.1626960	1
.5817731	.7272172	2.3010261	9.8616641	2
1.3089593	2.0361765	2.4771063	0.3088154	3
2.3268960	4.3630725	2.6020194	0.6397924	4
3.6353009	7.9983734	2.6988800	0.9030017	5
5.233596	13.231969	2.7779771	1.1216244	6
7.120730	20.352699	2.8447911	1.3086220	7
9.294991	29.647690	2.9025862	1.4719909	8
11.75374	41.40143	2.9534598	1.6170153	9
14.49319	55.89462	2.9988361	1.7473701	10
17.50803	73.40265	3.0397231	1.8657117	11
20.79117	94.19382	3.0768567	1.9740224	12
24.33329	118.52711	3.1107877	2.0738177	13
28.12251	146.64962	3.1419362	2.1662811	14
32.14395	178.79357	3.1706269	2.2523519	15
36.37932	215.17289	3.1971131	2.3327875	16
40.80649	255.97938	3.2215938	2.4082049	17
45.39905	301.37843	3.2442250	2.4791121	18
50.12591	351.50434	3.2651291	2.5459307	19
54.95090	406.45524	3.2844009	2.6090128	20

Point	Log $\frac{x}{y}$ =	Deflection angle,		
n.	log tan *i*.	*i*.		
11	8.8259886	3° 49′ 56.″39		
12	8.8971657	4° 30′ 43.″95		
13	8.9630300	5° 14′ 50.″28		
14	9.0243449	6° 02′ 14.″93		
15	9.0817250	6° 52′ 57.″31		
16	9.1356744	7° 46′ 56.″71		
17	9.1866111	8° 44′ 12.″26		
18	9.2348871	9° 44′ 42.″92		
19	9.2808016	10° 48′ 27.″44		
20	9.3246119	11° 55′ 24.″34		

TABLE II.

DEFLECTION ANGLES, FOR LOCATING SPIRAL CURVES IN THE
FIELD.

Rule for finding a Deflection.

Read under the *heading* corresponding to the point at which the instrument stands, and on the *line* of the number of the point observed.

	INSTRUMENT AT S. $s = 0.$	
No. of Point, n.	Deflection from Tangent, i.	Difference of Deflection.
0	00′	05′
1	05	07 30″
2	12 30″	10 50
3	23 20	14 10
4	37 30	17 30
5	55 00	20 50
6	1° 15 50	24 10
7	1 40 00	27 29
8	2 07 29	30 50
9	2 38 19	34 09
10	3 12 28	37 28
11	3 49 56	40 48
12	4 30 44	44 06
13	5 14 50	47 25
14	6 02 15	50 42
15	6 52 57	54 00
16	7 46 57	57 15
17	8 44 12	60 31
18	9 44 43	63 44
19	10 48 27	66 57
20	11 55 24	

TABLE II.—Deflection Angles.

Inst. at 1. $s = 0° 10'$.

No. of Point.	Deflection from aux. tan.	Diff. of Deflection.
0	05'	05'
1	00	10
2	10	12 30"
3	22 30"	15 50
4	38 20	19 10
5	57 30	22 30
6	1° 20 00	25 50
7	1 45 50	29 10
8	2 15 00	32 29
9	2 47 29	35 49
10	3 23 18	39 09
11	4 02 27	42 28
12	4 44 55	45 47
13	5 30 42	49 05
14	6 19 47	52 24
15	7 12 11	55 40
16	8 07 51	58 58
17	9 06 49	62 12
18	10 09 01	65 27
19	11 14 28	68 40
20	12 23 08	

Inst. at 2. $s = 0° 30'$.

No. of Point.	Deflection from aux. tan.	Diff. of Deflection.
0	17' 30"	7' 30"
1	10	10
2	00	15
3	15	17 30
4	32 30	20 50
5	53 20	24 10
6	1° 17 30	27 30
7	1 45 00	30 50
8	2 15 50	34 09
9	2 49 59	37 30
10	3 27 29	40 49
11	4 08 18	44 08
12	4 52 26	47 28
13	5 39 54	50 46
14	6 30 40	54 04
15	7 24 44	57 22
16	8 22 06	60 39
17	9 22 45	63 54
18	10 26 39	67 10
19	11 33 49	70 23
20	12 44 12	

Inst. at 3. $s = 1° 00'$.

No. of Point.	Deflection from aux. tan.	Diff. of Deflection.
0	36' 40"	9' 10"
1	27 30	12 30
2	15	15
3	00	20
4	20	22 30
5	42 30	25 50
6	1° 08 20	29 10
7	1 37 30	32 30
8	2 10 00	35 50
9	2 45 50	39 09
10	3 24 59	42 29
11	4 07 28	45 49
12	4 53 17	49 08
13	5 42 25	52 27
14	6 34 52	55 45
15	7 30 37	59 03
16	8 29 40	62 21
17	9. 32 01	65 36
18	10 37 37	68 52
19	11 46 29	72 06
20	12 58 35	

Inst. at 4. $s = 1° 40'$.

No. of Point.	Deflection from aux. tan.	Diff. of Deflection.
0	1° 02' 30"	10' 50"
1	51 40	14 10
2	37 30	17 30
3	20	20
4	00	25
5	25	27 30
6	52 30	30 50
7	1 23 20	34 10
8	1 57 30	37 30
9	2 35 00	40 50
10	3 15 50	44 09
11	3 59 59	47 29
12	4 47 28	50 48
13	5 38 16	54 08
14	6 32 24	57 26
15	7 29 50	60 44
16	8 30 34	64 02
17	9 34 36	67 19
18	10 41 55	70 34
19	11 52 29	73 49
20	13 06 18	

TABLE II.—DEFLECTION ANGLES.

INST. AT 5. $s = 2° 30'$.			INST. AT 6. $s = 3° 30'$.		
No. of Point.	Deflection from aux. tan.	Diff. of Deflection.	No. of Point.	Deflection from aux. tan.	Diff. of Deflection.
0	1° 35′ 00″		0	2° 14′ 10″	
1	1 22 30	.12′ 30″	1	2 00 00	14′ 10″
2	1 06 40	15 50	2	1 42 30	17 30
3	47 30	19 10	3	1 21 40	20 50
4	25	22 30	4	57 30	24 10
5	00	25	5	30	27 30
6	30	30	6	00	30
7	1 02 30	32 30	7	35	35
8	1 38 20	35 50	8	1 12 30	37 30
9	2 17 30	39 10	9	1 53 20	40 50
10	3 00 00	42 30	10	2 37 30	44 10
11	3 45 50	45 50	11	3 25 00	47 30
12	4 34 59	49 09	12	4 15 49	50 49
13	5 27 28	52 29	13	5 09 58	54 09
14	6 23 15	55 47	14	6 07 27	57 29
15	7 22 23	59 08	15	7 08 15	60 48
16	8 24 48	62 25	16	8 12 21	64 06
17	9 30 31	65 43	17	9 19 46	67 25
18	10 39 32	69 01	18	10 30 28	70 42
19	11 51 48	72 16	19	11 44 27	73 59
20	13 07 20	75 32	20	13 01 41	77 14

INST. AT 7. $s = 4° 40'$.			INST. AT 8. $s = 6° 00'$.		
No. of Point.	Deflection from aux. tan.	Diff. of Deflection.	No. of Point.	Deflection from aux. tan.	Diff. of Deflection.
0	3° 00′ 00″		0	3° 52′ 31″	
1	2 44 10	15′ 50″	1	3 35 00	17′ 31″
2	2 25 00	19 10	2	3 14 10	20 50
3	2 02 30	22 30	3	2 50 00	24 10
4	1 36 40	25 50	4	2 22 30	27 30
5	1 07 30	29 10	5	1 51 40	30 50
6	35	32 30	6	1 17 30	34 10
7	00	35	7	40	37 30
8	40	40	8	00	40
9	1 22 30	42 30	9	45	45
10	2 08 20	45 50	10	1 32 30	47 30
11	2 57 30	49 10	11	2 23 20	50 50
12	3 50 00	52 30	12	3 17 30	54 10
13	4 45 49	55 49	13	4 15 00	57 30
14	5 44 58	59 09	14	5 15 49	60 49
15	6 47 26	62 28	15	6 19 58	64 09
16	7 53 14	65 48	16	7 27 26	67 28
17	9 02 19	69 05	17	8 38 13	70 47
18	10 14 43	72 24	18	9 52 18	74 05
19	11 30 24	75 41	19	11 09 40	77 22
20	12 49 21	78 57	20	12 30 20	80 40

TABLE II.—DEFLECTION ANGLES.

No. of Point	INST. AT 9. $s = 7° 30'$. Deflection from aux. tan.	Diff. of Deflection.
0	4° 51' 41"	
1	4 32 31	19' 10"
2	4 10 01	22 30
3	3 44 10	25 51
4	3 15 00	29 10
5	2 42 30	32 30
6	2 06 40	35 50
7	1 27 30	39 10
8	45	42 30
9	00	45
10	50	50
11	1 42 30	52 30
12	2 38 20	55 50
13	3 37 30	59 10
14	4 40 00	62 30
15	5 45 49	65 49
16	6 54 57	69 08
17	8 07 25	72 28
18	9 23 11	75 46
19	10 42 16	79 05
20	12 04 38	82 22

No. of Point	INST. AT 10. $s = 9° 10'$. Deflection from aux. tan.	Diff. of Deflection.
0	5° 57' 32"	
1	5 36 42	20' 50"
2	5 12 31	24 11
3	4 45 01	27 30
4	4 14 10	30 51
5	3 40 00	34 10
6	3 02 30	37 30
7	2 21 40	40 50
8	1 37 30	44 10
9	50	47 30
10	00	50
11	55	55
12	1 52 30	57 30
13	·2 53 20	60 50
14	3 57 30	64 10
15	.5 05 00	67 30
16	6 15 49	70 49
17	7 29 57	74 08
18	8 47 24	77 27
19	10 08 10	80 46
20	11 32 14	84 04

No. of Point	INST. AT 11. $s = 11° 00'$. Deflection from aux. tan.	Diff. of Deflection.
0	7° 10' 04"	
1	6 47 33	22' 31"
2	6 21 42	25 51
3	5 52 32	29 10
4	5 20 01	32 31
5	4 44 10	35 51
6	4 05 00	39 10
7	3 22 30	42 30
8	2 36 40	45 50
9	1 47 30	49 10
10	55	52 30
11	00	55
12	1 00 00	60
13	2 02 30	62 30
14	3 08 20	65 50
15	4 17 30	69 10
16	5 29 59	72 30
17	6 45 48	75 49
18	8 04 57	79 09
19	9 27 24	82 27
20	10 53 09	85 45

No. of Point	INST. AT 12. $s = 13° 00'$. Deflection from aux. tan.	Diff. of Deflection.
0	8° 29' 16"	
1	8 05 05	24' 11"
2	7 37 34	27 31
3	7 06 43	30 51
4	6 32 32	34 11
5	5 55 01	37 31
6	5 14 11	40 50
7	4 30 00	44 11
8	3 42 30	47 30
9	2 51 40	50 50
10	1 57 30	54 10
11	1 00 00	57 30
12	00	60
13	1 05 00	65 ·
14	2 12 30	67 30
15	3 23 20	70 50
16	4 37 30	74 10
17	5 54 59	77 29
18	7 15 48	80 49
19	8 39 56	84 08
20	10 07 23	87 27

TABLE II.—Deflection Angles.

No. of Point.	INST. AT 13. s = 15° 10'. Deflection from aux. tan.	Diff. of Deflection.	No. of Point.	INST. AT 14. s = 17° 30'. Deflection from aux. tan.	Diff. of Deflection.
0	9° 55' 10"		0	11° 27' 45"	
		25' 52"			27' 32"
1	9 29 18		1	11 00 13	
		29 12			30 53
2	9 00 06		2	10 29 20	
		32 31			34 12
3	8 27 35		3	9 55 08	
		35 51			37 32
4	7 51 44		4	9 17 36	
		39 12			40 51
5	7 12 32		5	8 36 45	
		42 30			44 12
6	6 30 02		6	7 52 33	
		45 51			47 31
7	5 44 11		7	7 05 02	
		49 11			50 51
8	4 55 00		8	6 14 11	
		52 30			54 11
9	4 02 30		9	5 20 00	
		55 50			57 30
10	3 06 40		10	4 22 30	
		59 10			60 50
11	2 07 30		11	3 21 40	
		62 30			64 10
12	1 05 00		12	2 17 30	
		65			67 30
13	00		13	1 10 00	
		70			70
14	1 10 00		14	00	
		72 30			75
15	2 22 30		15	1 15 00	
		75 50			77 30
16	3 38 20		16	2 32 30	
		79 10			80 50
17	4 57 30		17	3 53 20	
		82 29			84 10
18	6 19 59		18	5 17 30	
		85 49			87 29
19	7 45 48		19	6 44 59	
		89 08			90 49
20	9 14 56		20	8 15 48	

No. of Point.	INST. AT 15. s = 20° 00'. Deflection from aux. tan.	Diff. of Deflection.	No. of Point.	INST. AT 16. s = 22° 40'. Deflection from aux. tan.	Diff. of Deflection.
0	13° 07' 03"		0	14° 53' 03"	
		29' 14"			30' 54"
1	12 37 49		1	14 22 09	
		32 33			34 15
2	12 05 16		2	13 47 54	
		35 53			37 34
3	11 29 23		3	13 10 20	
		39 13			40 54
4	10 50 10		4	12 29 26	
		42 33			44 14
5	10 07 37		5	11 45 12	
		45 52			47 33
6	9 21 45		6	10 57 39	
		49 11			50 53
7	8 32 34		7	10 06 46	
		52 32			54 12
8	7 40 02		8	9 12 34	
		55 51			57 31
9	6 44 11		9	8 15 03	
		59 10			60 52
10	5 45 01		10	6 14 11	
		62 31			64 10
11	4 42 30		11	6 10 01	
		65 50			67 31
12	3 36 40		12	5 02 30	
		69 10			70 50
13	2 37 30		13	3 51 40	
		72 30			74 10
14	1 15 00		14	2 37 30	
		75			77 30
15	00		15	1 20 00	
		80			80
16	1 20 00		16	00	
		82 30			85
17	2 42 30		17	1 25 00	
		85 50			87 30
18	4 08 20		18	2 52 30	
		89 10			90 50
19	5 37 30		19	4 23 20	
		92 29			94 10
20	7 09 59		20	5 57 30	

TABLE II.—DEFLECTION ANGLES.

INST. AT 17. s = 25° 30'.

No. of Point.	Deflection from aux. tan.	Diff. of Deflection.
0	16° 45' 48'	32' 37"
1	16 13 11	36 56
2	15 37 15	39 16
3	14 57 59	42 35
4	14 15 24	45 55
5	13 29 29	49 15
6	12 40 14	52 33
7	11 47 41	55 54
8	10 51 47	59 12
9	9 52 35	62 32
10	8 50 03	65 51
11	7 44 12	69 11
12	6 35 01	72 31
13	5 22 30	75 50
14	4 06 40	79 10
15	2 47 30	82 30
16	1 25 00	85
17	00	90
18	1 30 00	92 30
19	3 02 30	95 50
20	4 38 20	

INST. AT 18. s = 28° 30'.

No. of Point.	Deflection from aux. tan.	Diff. of Deflection.
0	18° 45' 17"	34' 18"
1	18 10 59	37 38
2	17 33 21	40 58
3	16 52 23	44 18
4	16 08 05	47 37
5	15 20 28	50 56
6	14 29 32	54 15
7	13 35 17	57 35
8	12 37 42	60 53
9	11 36 49	64 13
10	10 32 36	67 33
11	9 25 03	70 51
12	8 14 12	74 11
13	7 00 01	77 31
14	5 42 30	80 50
15	4 21 40	84 10
16	2 57 30	87 30.
17	1 30 00	90
18	. 00	95
19	1 35 00	97 30
20	3 12 30	

INST. AT 19. s = 31° 40'.

No. of Point.	Deflection from aux. tan.	Diff. of Deflection.
0	20° 51' 33"	36' 01"
1	20 15 32	39 21
2	19 36 11	42 40
3	18 53 31	46 00
4	18 07 31	49 19
5	17 18 12	52 39
6	16 25 33	55 57
7	15 29 36	59 16
8	14 30 20	62 36
9	13 27 44	65 54
10	12 21 50	69 14
11	11 12 36	75 32
12	10 00 04	75 52
13	8 44 12	79 11
14	7 25 01	82 31
15	6 02 30	85 50
16	4 36 40	89 10
17	3 07 30	92 30
18	1 35	95
19	00	100
20	1 40	

INST. AT 20. s = 35° 00'.

No. of Point.	Deflection from aux. tan.	Diff. of Deflection.
0	23° 04' 36"	37' 44"
1	22 26 52	41 04
2	21 45 48	44 23
3	21 01 25	47 43
4	20 13 42	51 02
5	19 22 40	54 21
6	18 28 19	57 40
7	17 30 39	60 59
8	16 29 40	64 17
9	15 25 23	67 37
10	14 17 46	70 55
11	13 06 51	74 14
12	11 52 37	77 33
13	10 35 04	80 52
14	9 14 12	84 11
15	7 50 01	87 31
16	6 22 30	90 50
17	4 51 40	94 10
18	3 17 30	97 30
19	1 40	100
20	00	

TABLE III.

DEGREE OF CURVE AND VALUES OF THE · COORDINATES x AND y, FOR EACH CHORD-POINT OF THE SPIRAL FOR VARIOUS LENGTHS OF THE CHORD.

			$c.$ CHORD-LENGTH = 10.		
$n.$	$nc.$	$Ds.$	$y.$	$x.$	Log $x.$
1	10	1° 40′ 00″	10.000	0.0145	8.162696
2	20	3 20 02	20.000	.0727	8.861664
3	30	5 00 06	29.999	.2036	9.308815
4	40	6 40 13	39.996	.4363	9.639792
5	50	8 20 26	49.990	.7998	9.903002
6	60	10 00 45	59.976	1.323	0.121624
7	70	11 41 12	69.951	2.035	0.308622
8	80	13 21 48	79.907	2.965	0.471991
9	90	15 02 34	89.838	4.140	0.617015
10	100	16 43 31	99.732	5.589	0.747370
11	110	18 24 42	109 578	7.340	0.805712
12	120	20 06 07	119.359	9.419	0.974022
13	130	21 47 48	129.059	11.853	1.073818
14	140	23 29 46	138.655	14.665	1.166281
15	150	25 12 02	148.125	17.879	1.252352
16	160	26 54 39	157.439	21.517	1.332788
17	170	28 37 38	166.569	25.598	1.408205
18	180	30 21 01	175.479	30.138	1.479112
19	190	32 04 48	184.132	35.150	1.545931
20	200	33 49 02	192.487	40.645	1.609013
		35 33 46			

58

TABLE III.

c. CHORD-LENGTH = 11.

n.	nc.	Ds.	y.	x.	Log x.
1	11	1° 30′ 55″	11.000	0.0160	8.204089
2	22	3 01 50	22.000	.0800	8.903057
3	33	4 32 48	32.999	.2240	9.350208
4	44	6 03 48	43.996	.4799	9.681185
5	55	7 34 52	54.989	.8798	9.944394
6	66	9 06 01	65.974	1.456	0.163017
7	77	10 37 16	76.946	2.239	0.350015
8	88	12 08 37	87.898	3.261	0.513384
9	99	13 40 06	98.822	4.554	0.658408
10	110	15 11 44	109.706	6.148	0.788763
11	121	16 43 31	120.536	8.074	0.907104
12	132	18 15 29	131.295	10.361	1.015415
13	143	19 47 39	141.965	13.038	1.115210
14	154	21 20 01	152.521	16.131	1.207674
15	165	22 52 38	162.937	19.667	1.293745
16	176	24 25 29	173.183	23.669	1.374180
17	187	25 58 36	183.226	28.158	1.449598
18	198	27 32 01	193.027	33.152	1.520505
19	209	29 05 45	202.545	38.665	1.587323
20	220	30 39 48	211.735	44.710	1.650405
		32 14 11			

c. CHORD-LENGTH = 12.

n.	nc.	Ds.	y.	x.	Log x.
1	12	1° 23′ 20″	12.000	0.0175	8.241877
2	24	2 46 41	24.000	.0873	8.940845
3	36	4 10 03	35.999	.2443	9.387997
4	48	5 33 28	47.996	.5236	9.718974
5	60	6 56 55	59.988	.9598	9.982183
6	72	8 20 26	71.971	1.588	0.200806
7	84	9 44 01	83.941	2.442	0.387803
8	96	11 07 42	95.889	3.558	0.551172
9	108	12 31 28	107.806	4.968	0.696196
10	120	13 55 21	119.679	6.707	0.826551
11	132	15 19 22	131.493	8.808	0.944893
12	144	16 43 31	143.231	11.303	1.053204
13	156	18 07 48	154.871	14.223	1.152999
14	168	19 32 15	166.386	17.598	1.245462
15	180	20 56 53	177.749	21.455	1.331533
16	192	22 21 43	188.927	25.821	1.411969
17	204	23 46 44	199.883	30.718	1.487386
18	216	25 11 59	210.575	36.165	1.558293
19	228	26 37 28	220.958	42.181	1.625113
20	240	28 03 12	230.984	48.774	1.688194
		29 29 12			

TABLE III.

c. CHORD-LENGTH = 13.

n.	nc.	Ds.	y.	x.	Log x.
1	13	1° 16′ 55″	13.000	0.0189	8.276639
2	26	2 33 52	26.000	.0945	8.975607
3	39	3 50 49	38.999	.2647	9 422759
4	52	5 07 48	51.995	.5672	9.753736
5	65	6 24 49	64.987	1.040	0.016945
6	78	7 41 53	77.969	1.720	0.235568
7	91	8 59 00	90.936	2.646	0.422565
8	104	10 16 12	103.879	3.854	0.585934
9	117	11 33 28	116.789	5.382	0.730959
10	130	12 50 49	129.652	7.266	0.861313
11	143	14 08 16	142.451	9.542	0.979655
12	156	15 25 50	155.167	12.245	1.087966
13	169	16 43 30	167.776	15.409	1.187761
14	182	18 01 18	180.252	19.064	1.280224
15	195	19 19 14	192.562	23.243	1.366295
16	208	20 37 20	204.671	27.972	1.446731
17	221	21 55 34	216.540	33.277	1.522148
18	234	23 14 00	228.123	39.179	1.593055
19	247	24 32 35	239.371	45.696	1.659874
20	260	25 51 23	250.233	52.839	1.722956.
		27 10 23			

c. CHORD-LENGTH = 14.

n.	nc.	Ds	y.	x.	Log x.
1	14	1° 11′ 26″	14.000	0.0204	8.308824
2	28	2 22 52	28.000	.1018	9.007792
3	42	3 34 19	41.999	.2851	9.454943
4	56	4 45 48	55.995	.6108	9.785920
5	70	5 57 18	69.986	1.120	0.049130
6	84	7 08 51	83.966	1.852	0.267752
7	98	8 20 26	97.931	2.849	0.454750
8	112	9 32 04	111.870	4.151	0.618119
9	126	10 43 47	125.773	5.796	0.763143
10	140	11 55 33	139.625	7.825	0.893498
11	154	13 07 24	153.409	10.276	1.011840
12	168	14 19 20	167.103	13.187	1.120150
13	182	15 31 22	180.682	16.594	1.219946
14	196	16 43 29	194.117	20.531	1.312409
15	210	17 55 44	207.374	25.031	1.398480
16	224	19 06 05	220.415	30.124	1.478915
17	238	20 20 34	233.196	35.837	1.554333
18	252	21 33 11	245.670	42.193	1.625240
19	266	22 45 56	257.785	49.211	1.692059
20	280	23 58 51	269.481	56.903	1.755141
		25 11 55			

TABLE III.

c. CHORD-LENGTH = 15.

n.	nc.	Ds.	y.	x.	Log x.
1	15	1° 06′ 40″	15.000	0.0218	8.338787
2	30	2 13 20	30.000	.1091	9.037755
3	45	3 20 02	44.998	.3054	9.484907
4	60	4 26 44	59.994	.6545	9.815884
5	75	5 33 28	74.984	1.200	0.079093
6	90	6 40 13	89.964	1.985	0.297716
7	105	7 47 01	104.926	3.053	0.484713
8	120	8 53 51	119.861	4.447	0.648082
9	135	10 00 45	134.757	6.216	0.793107
10	150	11 07 41	149.599	8.384	0.923461
11	165	12 14 41	164.367	11.010	1.041803
12	180	13 21 47	179.039	14.129	1.150114
13	195	14 28 56	193.588	17.779	1.249909
14	210	15 36 09	207.983	21.997	1.342372
15	225	16 43 28	222.187	26.819	1.428443
16	240	17 50 54	236.159	32.276	1.508879
17	255	18 58 25	249.853	38.397	1.584296
18	270	20 06 02	263 218	45.207	1.655203
19	285	21 13 47	276.198	52.726	1.722022
20	300	22 21 39	288.730	60.968	1.785104
		23 29 48			

c. CHORD-LENGTH = 16.

n.	nc	Ds.	y.	x.	Log x.
1	16	1° 02′ 30″	16 000	0.0233	8.366816
2	32	2 05 00	32.000	.1164	9.065784
3	48	3 07 31	47.998	.3258	9.512935
4	64	4 10 03	63.994	.6981	9.843912
5	80	5 12 36	79.983	1.230	0.107122
6	96	6 15 11	95.961	2.117	0.325744
7	112	7 17 47	111.921	3.256	0.512742
8	128	8 20 26	127.852	4.744	0.676111
9	144	9 23 07	143.741	6.624	0.821135
10	160	10 25 51	159.572	8.943	0.951490
11	176	11 28 37	175.325	11.744	1.069832
12	192	12 31 28	190.975	15.071	1.178142
13	208	13 34 21	206.494	18.964	1.277938
14	224	14 37 20	221.848	23.464	1.370401
15	240	15 40 21	236.999	28.607	1.456472
16	256	16 43 28	251.903	34.428	1.536927
17	272	17 46 40	266.510	40.957	1.612325
18	288	18 49 57	280.766	48.221	1.683232
19	304	19 53 20	294.611	56.241	1.750051
20	320	20 56 49	307.979	65.032	1.813133
		22 00 23			

TABLE III.

c. CHORD-LENGTH = 17.

n.	*nc.*	*Ds.*	*y.*	*x.*	Log *x.*
1	17	0° 58′ 49″	17.000	0.0247	8.393145
2	34	1 57 38	34.000	.1236	9.092113
3	51	2 56 27	50.998	.3461	9.539264
4	68	3 55 19	67.994	.7417	9.870241
5	85	4 54 12	84.982	1.360	0.133451
6	102	5 53 06	101.959	2.249	0.352073
7	119	6 52 00	118.916	3.460	0.539071
8	136	7 50 57	135.842	5.040	0.702440
9	153	8 49 55	152.725	7.038	0.847464
10	170	9 48 56	169.545	9.502	0.977819
11	187	10 48 00	186.282	12.478	1.096161
12	204	11 47 07	202.911	16.013	1.204471
13	221	12 46 15	219.400	20.150	1.304267
14	238	13 45 27	235.714	24.930	1.396730
15	255	14 44 44	251.812	30.395	1.482801
16	272	15 44 03	267.647	36.579	1.563236
17	289	16 43 27	283.167	43.516	1.638654
18	306	17 42 56	298.314	51.234	1.709561
19	323	18 42 29	313.024	59.756	1.776380
20	340	19 42 07	327.228	69.097	1.839462
		20 41 49			

c. CHORD-LENGTH = 18.

n.	*nc.*	*Ds.*	*y.*	*x.*	Log *x.*
1	18	0° 55′ 33″	18.000	0.0262	8.417968
2	36	1 51 07	36.000	.1309	9.116937
3	54	2 46 40	53.998	.3665	9.564088
4	72	3 42 16	71.993	.7853	9.895065
5	90	4 37 51	89.981	1.440	0.158274
6	108	5 33 28	107.957	2.382	0.376897
7	126	6 29 05	125.911	3.663	0.563894
8	144	7 24 45	143.833	5.337	0.727263
9	162	8 20 26	161.708	7.452	0.872288
10	180	9 16 08	179.518	10.061	1.002643
11	198	10 11 54	197.240	13.212	1.120984
12	216	11 07 41	214.847	16.955	1.229295
13	234	12 03 31	232.306	21.335	1.329090
14	252	12 59 24	249.579	26.397	1.421554
15	270	13 55 20	266.624	32.183	1.507624
16	288	14 51 18	283.391	38.731	1.588060
17	306	15 47 20	299.824	46.076	1.663477
18	324	16 43 27	315.862	54.248	1.734385
19	342	17 39 37	331.437	63.271	1.801203
20	360	18 35 51	346.476	73.161	1.864285
		19 32 08			

TABLE III.

c. CHORD-LENGTH = 19.

n.	nc.	Ds.	y.	x.	Log x.
1	19	0° 52′ 38″	19.000	0.0276	8.441450
2	38	1 45 16	38.000	.1382	9.140418
3	57	2 37 54	56.998	.3869	9.587569
4	76	3 30 34	75.993	.8290	9.918546
5	95	4 23 13	94.980	1.520	0.181755
6	114	5 15 54	113.954	2.514	0.400378
7	133	6 08 36	132.906	3.867	0.587376
8	152	7 01 19	151.824	5.633	0.750744
9	171	7 54 03	170.692	7.866	0.895769
10	190	8 46 49	189.491	10.620	1.026124
11	209	9 39 36	208.198	13.947	1.144465
12	228	10 32 26	226.783	17.897	1.252776
13	247	11 25 18	245.212	22.520	1.352571
14	266	12 18 12	263.445	27.863	1.445035
15	285	13 11 09	281.437	33.971	1.531105
16	304	14 04 09	299.135	40.883	1.611541
17	323	14 57 11	316.481	48.636	1.686958
18	342	15 50 16	333.410	57.262	1.757866
19	361	16 43 25	349.851	66.786	1.824684
20	380	17 36 38	365.725	77.226	1.887766
		18 29 54			

c. CHORD-LENGTH = 20.

n.	nc.	Ds.	y.	x.	Log x.
1	20	0° 50′ 00″	20.000	0.0291	8.463726
2	40	1 40 00	40.000	.1454	9.162694
3	60	2 30 01	59.998	.4072	9.609845
4	80	3 20 02	79.993	.8726	9.940822
5	100	4 10 03	99.979	1.600	0.204032
6	120	5 00 05	119.952	2.646	0.422654
7	140	5 50 08	139.901	4.071	0.609652
8	160	6 40 13	159.815	5.930	0.773021
9	180	7 30 18	179.676	8.280	0.918045
10	200	8 20 26	199.465	11.179	1.048400
11	220	9 10 34	219.156	14.681	1.166742
12	240	10 00 44	238.719	18.839	1.275052
13	260	10 50 56	258.118	23.705	1.374848
14	280	11 41 10	277.310	29.330	1.467311
15	300	12 31 26	296.249	35.759	1.553382
16	320	13 21 45	314.879	43.035	1.633817
17	340	14 12 06	333.138	51.196	1.709235
18	360	15 02 29	350.958	60.276	1.780142
19	380	15 52 55	368.264	70.301	1.846961
20	400	16 43 25	384.974	81.290	1.910043
		17 33 58			

TABLE III.

c. CHORD-LENGTH = 21.

n.	nc.	Ds.			y.	x.	Log. x.
1	21	0°	47′	37″	21.000	0.0305	8.484915
2	42	1	35	.14	42.000	.1527	9.183883
3	63	2	22	52	62.998	.4276	9.631035
4	84	3	10	30	83.992	.9162	9.962012
5	105	3	58	08	104.978	1.680	0.225221
6	126	4	45	47.	125.949	2.779	0.443844
7	147	5	33	27	146.896	4.274	0.630841
8	168	6	21	08	167.805	6.226	0.794210
9	189	7	08	50	188.660	8.694	0.939235
10	210	7	56	33	209.438	11.738	1.069589
11	231	8	44	18	230.114	15.415	1.187931
12	252	9	32	03	250.655	19.781	1.296242
13	273	10	19	51	271.023	24.891	1.396037
14	294	11	07	40	291.176	30.706	1.488500
15	315	11	55	31	311.062	37.547	1.574571
16	336	12	43	24	330.623	45.186	1.655007
17	357	13	31	20	349.795	53.756	1.730424
18	378	14	19	17	368.506	63.289	1.801331
19	399	15	07	17	386.677	73.816	1.868150
		15	55	19			

c. CHORD-LENGTH = 22.

n.	nc.	Ds.			y.	x.	Log. x.
1	22		45′	27″	22.000	0.0320	8.505119
2	44	1°	30	53	44.000	.1600	9.204087
3	66	2	16	22	65.998	.4480	9.651238
4	88	3	01	50	87.992	.9599	9.982215
5	110	3	47	18	109.977	1.760	0.245424
6	132	4	32	48	131.947	2.911	0.464047
7	154	5	18	18	153.891	4.478	0.651045
8	176	6	03	48	175.796	6.522	0.814414
9	198	6	49	19	197.643	9.108	0.959438
10	220	7	34	51	219.411	12.297	1.089793
11	242	8	20	25	241.071	16.149	1.208134
12	264	9	06	00	262.591	20.723	1.316445
13	286	9	51	36	283.929	26.076	1.416240
14	308	10	37	13	305.042	32.263	1.508704
15	330	11	22	53	325.874	39.335	1.594775
16	352	12	08	34	346.367	47.338	1.675210
17	374	12	54	16	366.451	56.315	1.750623
18	396	13	40	01	386.054	66.303	1.821535
		14	25	49			

TABLE III.

c. CHORD-LENGTH = 23.

n.	*nc.*	*D_s.*	*y.*	*x.*	Log. *x.*
1	23	0° 43′ 29″	23.000	0.0335	8.524424
2	46	1 26 58	46.000	.1673	9.223392
3	69	2 10 26	68.998	.4683	9.670543
4	92	2 53 56	91.991	1.004	0.001520
5	115	3 37 26	114.976	.1.840	0.264729
6	138	4 20 56	137.945	3.043	0.483352
7	161	5 04 26	160.886	4.681	0.670350
8	184	5 47 58	183.787	6.819	0.833719
9	207	6 31 30	206.627	9.522	0.978743
10	230	7 15 04	229.384	12.856	1.109098
11	253	7 58 38	252.029	16.883	1.227439
12	276	8 42 13	274 527	21.665	1.335750
13	299	9 25 49	296.835	27.261	1.435545
14	322	10 09 27	318.907	33.729	1.528009
15	345	10 53 06	340.686	41.123	1.614080
16	368	11 36 47	362.110	49.490	1.694515
17	391	12 20 29	383.108	58.875	1.769933
		13 04 13			

c. CHORD-LENGTH = 24.

n.	*nc.*	*D_s.*	*y'.*	*x.*	Log. *x.*
1	24	41′ 40″	24.000	0.0349	8.542907
2	48	1° 23 20	48.000	.1745	9.241875
3	72	2 05 00	71.998	.4887	9.689027
4	96	2 46 41	95.991	1.047	0.020004
5	120	3 28 22	119.975	1.920	0.283213
6	144	4 10 03	143.942	3.176	0.501836
7	168	4 51 45	167.881	4.885	0.688833
8	192	5 33 28	191.777	7.115	0.852202
9	216	6 15 10	215.611	9.936	0.997226
10	240	6 56 54	239.358	13.415	1.127581
11	264	7 38 39	262.987	17.617	1.245923
12	288	8 20 25	286.463	22.607	1.354234
13	312	9 02 12	309.741	28.446	1.454029
14	336	9 44 00	332.773	35.190	1.546492
15	360	10 25 48	355.499	42.910	1.632563
16	384	11 07 39	377.854	51.641	1.712999
17	408	11 49 31	399.765	61.435	1.788416
		12 31 25			

TABLE III.

c. CHORD-LENGTH = 25.

n.	nc.	Ds.	y.	x.	Log. x.
1	25	0° 40′ 00″	25.000	0.0364	8.560636
2	50	1 20 00	50.000	.1818	9.259604
3	75	2 00 00	74.997	.5090	9.706755
4	100	2 40 01	99.991	1.091	0.037732
5	125	3 20 02	124.974	2.000	0.300942
6	150	4 00 03	149.940	3.308	0.519564
7	175	4 40 04	174.876	5.088	0.706562
8	200	5 20 06	199.768	7.412	0.869931
9	225	6 00 09	224.595	10.350	1.014955
10	250	6 40 13	249.331	13.974	1.145310
11	275	7 20 17	273.945	18.351	1.263652
12	300	8 00 22	298.398	23.548	1.371962
13	325	8 40 28	322.647	29.632	1.471758
14	350	9 20 35	346.638	36.662	1.564221
15	375	10 00 43	370.311	44.698	1.650292
16	400	10 40 52	393.598	53.793	1.730727
		11 21 03			

c. CHORD-LENGTH = 26.

n.	nc.	Ds.	y.	x.	Log. x.
1	26	0° 38′ 28″	26.000	0.0378	8.577669
2	52	1 16 56	52.000	.1891	9.276637
3	78	1 55 24	77.997	.5294	9.723789
4	104	2 33 52	103.990	1.134	0.054766
5	130	3 12 20	129.973	2.080	0.317975
6	156	3 50 48	155.937	3.440	0.536598
7	182	4 29 18	181.871	5.292	0.723595
8	208	5 07 48	207.759	7.708	0.886964
9	234	5 46 18	233.579	10.764	1.031989
10	260	6 24 48	259.304	14.533	1.162343
11	286	7 03 20	284.903	19.085	1.280685
12	312	7 41 52	310.334	24.490	1.388996
13	338	8 20 25	335.553	30.817	1.488791
14	364	8 58 59	360.504	38.129	1.581254
15	390	9 37 33	385.124	46.486	1.667325
		10 16 09			

TABLE III.

c. CHORD-LENGTH = 27.

n.	nc.	D_s.	y.	x.	Log. x.
1	27	0° 37' 02"	27.000	0.0393	8.594060
2	54	1 14 04	54.000	.1963	9.293028
3	81	1 51 07	80.997	.5498	9.740179
4	108	2 28 10	107.990	1.178	0.071156
5	135	3 05 12	134.972	2.160	0.334365
6	162	3 42 15	161.935	3.573	0.552988
7	189	4 19 19	188.866	5.495	0.739986
8	216	4 56 23	215.750	8.005	0.903355
9	243	5 33 28	242.562	11.178	1.048379
10	270	6 10 32	269.277	15.092	1.178734
11	297	6 47 38	295.860	19.819	1.297075
12	324	7 24 44	322.270	25.432	1.405386
13	351	8 01 51	348.459	32.002	1.505181
14	378	8 38 59	374.369	39.595	1.597645
15	405	9 16 07	399.936	48.274	1.683716
		9 53 16			

c. CHORD-LENGTH = 28.

n.	nc.	$D_s^{\cdot}$.	y.	x.	Log. x.
1	28	0° 35' 42"	28.000	0.0407	8.609854
2	56	1 11 26	55.999	.2036	9.308822
3	84	1 47 08	83.997	.5701	9.755973
4	112	2 22 52	111.990	1.222	0.086950
5	140	2 58 36	139.971	2.240	0.350160
6	168	3 34 19	167.933	3.705	0.568782
7	196	4 10 03	195.862	5.699	0.755780
8	224	4 45 48	223.740	8.301	0.919149
9	252	5 21 32	251.546	11.592	1.064173
10	280	5 57 17	279.251	15.650	1.194528
11	308	6 33 03	306.818	20.553	1.312870
12	336	7 08 50	334.206	26.374	1.421180
13	364	7 44 36	361.365	33.188	1.520976
14	392	8 20 24	388.235	41.062	1.613439
		8 56 13			

TABLE III.

c. CHORD-LENGTH = 2(

n.	*nc.*	*D$_s$.*	*y.*	*x.*
1	29	0° 34′ 29″	29.000	0.04
2	58	1 08 58	57.999	.21
3	87	1 43 27	86.997	.59
4	116	2 17 56	115.989	1.26
5	145	2 52 26	144.970	2.32
6	174	3 26 55	173.930	3.83
7	203	4 01 26	202.857	5.90
8	232	4 35 56	231.731	8.59
9	261	5 10 26	260.530	12.00
10	290	5 44 57	289.224	16.20
11	319	6 19 29	317.776	21.28
12	348	6 54 01	346.142	27.31
13	377	7 28 34	374.271	34.37
14	406	8 03 07	402.100	42.52
		8 37 40		

c. CHORD-LENGTH = 3(

n.	*nc.*	*D$_s$.*	*y.*	*x.*
1	30	0° 33′ 20″	30.000	0.04
2	60	1 06 40	59.999	.21
3	90	1 40 00	89.997	.61
4	120	2 13 20	119.989	1.30
5	150	2 46 41	149.969	2.40
6	180	3 20 02	179.928	3.97
7	210	3 53 22	209.852	6.10
8	240	4 26 44	239.722	8.89
9	270	5 00 05	269.514	12.42
10	300	5 33 27	299.197	16.76
11	330	6 06 49	328.734	22.02
12	360	6 40 12	358.078	28.25
13	390	7 13 36	387.176	35.55
		7 47 00		

n.	nc.	Ls.	y.	x.	Log x.
1	31	0° 32′ 15″	31.000	0.0451	8.654058
2	62	1 04 31	61.999	.2254	9.353026
3	93	1 36 47	92.997	.6312	9.800177
4	124	2 09 02	123.988	1.353	0.131154
5	155	2 41 18	154.968	2.479	0.394363
6	186	3 13 34	185.925	4.102	0.612986
7	217	3 45 50	216.847	6.309	0.799984
8	248	4 18 07	247.713	9.191	0.963353
9	279	4 50 24	278.498	12.834	1.108377
10	310	5 22 41	309.170	17.327	1.238732
11	341	5 54 59	339.692	22.755	1.357073
12	372	6 27 17	370.014	29.200	1.465384
13	403	6 59 35	400.082	36.743	1.565179
		7 31 53			

CHORD-LENGTH = 32.

n.	nc.	Ds.	y.	x.	Log x.
1	32	0° 31′ 15″	32.000	0.0465	8.667846
2	64	1 02 30	63.999	.2327	9.366814
3	96	1 33 45	95.997	.6516	9.813965
4	128	2 05 00	127.988	1.396	0.144942
5	160	2 36 16	159.967	2.559	0.408152
6	192	3 07 31	191.923	4.234	0.626774
7	224	3 38 47	223.842	6.513	0.813772
8	256	4 10 03	255.703	9.487	0.977141
9	288	4 41 19	287.481	13.248	1.122165
10	320	5 12 36	319.144	17.886	1.252520
11	352	5 43 53	350.649	23.489	1.370802
12	384	6 15 10	381.950	30.142	1.479172
13	416	6 46 28	412.988	37.929	1.578968
		7 17 46			

TABLE III.

c. CHORD-LENGTH = 33.

n.	*nc.*	*Ds.*	*y.*	*x.*	Log. *x.*
1	33	0° 30′ 19″	33.000	0.0480	8.681210
2	66	1 00 36	65.999	.2400	9.380178
3	99	1 30 55	98.997	.6719	9.827329
4	132	2 01 13	131.988	1.440	0.158306
5	165	2 31 32	164.966	2.639	0.421516
6	198	3 01 50	197.921	4.367	0.640138
7	231	3 32 09	230.837	6.716	0.827136
8	264	4 02 28	263.694	9.784	0.990505
9	297	4 32 48	296.465	13.662	1.135529
10	330	5 03 07	329.117	18.445	1.265884
11	363	5 33 27	361.607	24.223	1.384226
12	396	6 03 47	393.886	31.084	1.492536
		6 34 07			

c. CHORD-LENGTH = 34.

n.	*nc.*	*Ds.*	*y.*	*x.*	Log. *x.*
1	34	0° 29′ 25″	34.000	0.0495	8.694175
2	68	0 58 49	67.999	.2473	9.393143
3	102	1 28 14	101.996	.6923	9.840294
4	136	1 57 39	135.987	1.483	0.171271
5	170	2 27 04	169.965	2.719	0.434481
6	204	2 56 29	203.918	4.499	0.653103
7	238	3 25 55	237.832	6 920	0.840101
8	272	3 55 20	271.685	10.080	1.003470
9	306	4 24 46	305.449	14.076	1.148494
10	340	4 54 12	339.090	19.004	1.278849
11	374	5 23 38	372.565	24.957	1.397191
12	408	5 53 05	405.822	32.026	1.505501
		6 22 11			

c. CHORD-LENGTH = 35.

n.	nc.	Ds.	y.	x.	Log x.
1	35	0° 28' 34"	35.000	0.0509	8.706764
2	70	0 57 09 ·	69.999	.2545	9.405732
3	105	1 25 43	104.996	.7127	9.852883
4	140	1 54 17.	139.987	1.527	0.183860
5	175	2 22 52	174.964	2.799	0.447070
6	210	2 51 27	209.916	4.631	0.665692
7	245	3 20 01	244.827	7.123	0.852690
8	280	3 48 36	279.675	10.377	1.016059
9	315	4 17 12	314.433	14.490	1.161083
10	350	4 45 47	349.063	19.563	1.291438
11	385	5 14 23	383.523	25.691	1.409780
12	420	5 43 00	417.758	32.968	1.518090
		6 09 36			

c. CHORD-LENGTH = 36.

n.	nc.	Ds.	y.	x.	Log x.
1	36	0° 27' 47"	36.000	0.0524	8.718998
2	72	0 55 33	71.999	.2618	9.417967
3	108	1 23 20	107.996	.7330	9.865118
4	144	1 51 07	143.987	1.571	0.196095
5	180	2 18 54	179.963	2.879	0.459304
6	216	2 46 41	215.913	4.764	0.677927
7	252	3 14 28	251.822	7.327	0.864924
8	288	3 42 15	287.666	10.673	1.028293
9	324	4 10 03	323.417	14.905	1.173318
10	360	4 37 51	359.037	20.122	1.303673
11	396	5 05 39	394.480	26.425	1.422014
		5 33 27			

c. CHORD-LENGTH = 37.

n.	nc.	D_s.	y'.	x.	Log x.
1	37	0° 27' 02''	37.000	0.0538	8.730898
2	74	0 54 03	73.999	.2691	9.429866
3	111	1 21 05	110.996	.7534	9.877017
4	148	1 48 07	147.986	1.614	0.207994
5	185	2 15 09	184.962	2.959	0.471203
6	222	2 42 11	221.911	4.896	0.689826
7	259	3 09 13	258.817	7.530	0.876824
8	296	3 36 15	295.657	10.970	1.040193
9	333	4 03 17	332.400	15.319	1.185217
10	370	4 30 20	369.010	20.681	1.315572
11	407	4 57 23	405.438	27.159	1.433913
		5 24 26			

.c. CHORD-LENGTH = 38.

n.	nc.	D_s.	y'.	x.	Log x.
1	38	0° 26' 19''	38.000	0.0553	8.742480
2	76	0 52 39	75.999	.2763	9.441448
3	114	1 18 57	113.996	.7737	9.888599
4	152	1 45 16	151.986	1.658	0.219576
5	190	2 11 35	189.961	3.039	0.482785
6	228	2 37 54	227.909	5.028	0.701408
7	266	3 04 14	265.812	7.734	0.888406
8	304	3 30 33	303.648	11.266	1.051774
9	342	3 56 53	341.384	15.733	1.196799
10	380	4 23 13	378.983	21.240	1.327154
11	418	4 49 33	416.396	27.893	1.445495
		5 15 53			

TABLE III.

c. CHORD-LENGTH = 39.

n.	*nc.*	*D_s.*	*y.*	*x.*	Log *x.*
1	39	0° 25′ 38″	39.000	0.0567	8.753761
2	78	0 51 17	77.999	.2836	9.452729
3	117	1 16 55	116.996	.7941	9.899880
4	156	1 42 34	155.985	1.702	0.230857
5	195	2 08 13	194.960	3.119	0.494066
6	234	2 33 51	233.906	5.160	0.712689
7	273	2 59 30	272.807	7.938	0.899687
8	312	3 25 09	311.638	11.563	1.063055
9	351	3 50 48	350.368	16.147	1.208080
10	390	4 16 28	388.956	21.799	1.338435
		4 42 07			

c. CHORD-LENGTH = 40.

n.	*nc.*	*D_s.*	*y.*	*x.*	Log *x.*
1	40	0° 25′ 00″	40.000	0.0582	8.764756
2	80	0 50 00	79.999	.2909	9.463724
3	120	1 15 00	119.996	.8145	9.910875
4	160	1 40 00	159.985	1.745	0.241852
5	200	2 05 00	199.959	3.199	0.505062
6	240	2 30 01	239.904	5.293	0.723684
7	280	2 55 01	279.802	8.141	0.910682
8	320	3 20 01	319.629	11.859	1.074051
9	360	3 45 02	359.352	16.561	1.219075
10	400	4 10 03	398.929	22.358	1.349430
		4 35 03			

c. CHORD-LENGTH = 41.

n.	*nc.*	*D_s.*	*y.*	*x.*	Log *x.*
1	41	0° 24′ 24″	41.000	0.0596	8.775480
2	82	0 48 47	81.999	.2982	9.474448
3	123	1 13 10	122.996	.8348	9.921599
4	164	1 37 34	163.985	1.789	0.252576
5	205	2 01 57	204.958	3.279	0.515786
6	246	2 26 21	245.901	5.425	0.734408
7	287	2 50 45	286.797	8.345	0.921406
8	328	3 15 09	327.620	12.156	1.084775
9	369	3 39 33	368.336	16.975	1.229799
10	410	4 03 57	408.903	22.917	1.360154
		4 28 21			

TABLE III.

c. CHORD-LENGTH = 42.

n.	*nc.*	*D_s.*	*y.*	*x.*	Log *x.*
1	42	0° 23′ 49″	42.000	0.0611	8.785945
2	84	0 47 37	83.999	.3054	9.484913
3	126	1 11 26	125.996	.8552	9.932065
4	168	1 35 14	167.984	1.832	0.263042
5	210	1 59 02	209.957	3.359	0.526251
6	252	2 22 52	251.899	5.557	0.744874
7	294	2 46 41	293.792	8.548	0.931871
8	336	3 10 30	335.611	12.452	1.095240
9	378	3 34 19	377.319	17.389	1.240265
10	420	3 58 08	418.876	23.476	1.370619
		4 21 57			

c. CHORD-LENGTH = 43.

n.	*nc.*	*D_s.*	*y.*	*x.*	Log *x.*
1	43	0° 23′ 15″	43.000	0.0625	8.796164
2	86	0 46 31	85.999	.3127	9.495133
3	129	1 09 46	128.996	.8755	9.942284
4	172	1 33 02	171.984	1.876	0.273261
5	215	1 56 17	214.955	3.439	0.536470
6	258	2 19 33	257.897	5.690	0.755093
7	301	2 42 48	300.787	8.752	0.942090
8	344	3 06 04	343.601	12.749	1.105459
9	387	3 29 20	386.303	17.803	1.250484
10	430	3 52 35	428.849	24.035	1.380839
		4 15 50			

c. CHORD-LENGTH = 44.

n.	*nc.*	*D_s.*	*y.*	*x.*	Log *x.*
1	44	0° 22′ 44″	44.000	0.0640	8.806149
2	88	0 45 27	87.999	.3200	9.505117
3	132	1 08 11	131.995	.8959	9.952268
4	176	1 30 55	175.984	1.920	0.283245
5	220	1 53 38	219.954	3.519	0.546454
6	264	2 16 22	263.894	5.822	0.765077
7	308	2 39 06	307.782	8.955	0.952075
8	352	3 01 50	351.592	13.045	1.115444
9	396	3 24 34	395.287	18.217	1.260468
		3 47 18			

TABLE III.

c. CHORD-LENGTH = 45.

n.	*nc.*	*Ds.*	*y.*	*x.*	Log *x.*
1	45	0° 22′ 13″	45.000	0.0655	8.815908
2	90	0 44 27	89.999	.3272	9.514877
3	135	1 06 40	134.995	.9163	9.962028
4	180	1 28 53	179.983	1.963	0.293005
5	225	1 51 07	224.953	3.599	0.556214
6	270	2 13 20	269.892	5.954	0.774837
7	315	2 35 34	314.778	9.159	0.961834
8	360	2 57 48	359.583	13.341	1.125203
9	405	3 20 01	404.271	18.631	1.270228
		3 42 15			

c. CHORD-LENGTH = 46.

n.	*nc.*	*Ds.*	*y.*	*x.*	Log *x.*
1	46	0° 21′ 44″	46.000	0.0669	8.825454
2	92	0 43 29	91.999	.3345	9.524422
3	138	1 05 13	137.995	.9366	9.971573
4	184	1 26 58	183.983	2.007	0.302550
5	230	1 48 42	229.952	3.679	0.565759
6	276	2 10 26	275.889	6.087	0.784382
7	322	2 32 11	321.773	9.362	0.971380
8	368	2 53 56	367.573	13.638	1.134749
9	414	3 15 40	413.255	19.045	1.279773
		3 37 24			

c. CHORD-LENGTH = 47.

n.	*nc.*	*Ds.*	*y.*	*x.*	Log *x.*
1	47	0° 21′ 16″	47.000	0.0684	8.834794
2	94	0 42 33	93.999	.3418	9.533762
3	141	1 03 50	140.995	.9570	9.980913
4	188	1 25 06	187.982	2.051	0.311890
5	235	1 46 23	234.951	3.759	0.575100
6	282	2 07 40	281.887	6.219	0.793722
7	329	2 28 57	·28.768	9.566	0.980720
8	376	2 50 14	375.564	13.934	1.144089
9	423	3 11 31	422.238	19.459	1.289113
		3 32 48			

TABLE III.

c. CHORD-LENGTH = 48.

n.	*nc.*	*Ds.*	*y.*	*x.*	Log *x.*
1	48	0° 20′ 50″	48.000	0.0698	8.843937
2	96	0 41 40	95.999	.3491	9.542905
3	144	1 02 30	143.995	.9774	9.990057
4	192	1 23 20	191.982	2.094	0.321034
5	240	1 44 10	239.950	3.839	0.584243
6	288	2 05 00	287.885	6.351	0.802866
7	336	2 25 51	335.763	9.769	0.989863
8	384	2 46 41	383.555	14.231	1.153232
		3 06 31			

c. CHORD-LENGTH = 49.

n.	*nc.*	*Ds.*	*y.*	*x.*	Log *x.*
1	49	0° 20′ 25″	49.000	0.0713	8.852892
2	98	0 40 49	97.999	.3563	9.551860
3	147	1 01 14	146.995	.9977	9.999011
4	196	1 21 38	195.982	2.138	0.329988
5	245	1 42 03	244.949	3.919	0.593198
6	294	2 02 27	293.882	6.484	0.811820
7	343	2 22 52	342.758	9.973	0.998818
8	392	2 43 17	391.546	14.527	1.162187
		3 03 31			

c. CHORD-LENGTH = 50.

n.	*nc.*	*Ds.*	*y.*	*x.*	Log *x.*
1	50	0° 20′ 00″	50.000	0.0727	8.861666
2	100	0 40 00	99.999	.3636	9.566634
3	150	1 00 00	149.995	1.018	0.007785
4	200	1 20 00	199.981	2.182	0.338762
5	250	1 40 00	249.948	3.999	0.601972
6	300	2 00 00	299.880	6.616	0.820594
7	350	2 20 00	349.753	10.176	1.007592
8	400	2 40 00	399.536	14.824	1.170961
		3 00 00			

76

TABLE IV.

Functions of the Angle *s*.

n.	*s.*	cos *s.*	log vers *s.*	$R\ 1° \times$ vers *s.*	sin *s.*	log sin *s.*	*s.*
I	0° 10′	.99999	4.626422	.024	.00291	7.463726	0° 10′
2	0 30	.99996	5.580662	.218	.00873	7.940842	0 30
3	I 00	.99985	6.182714	.873	.01745	8.241855	I 00
4	I 40	.99958	6.626392	2.424	.02908	8.463665	I 40
5	2 30	.99905	6.978536	5.453	.04362	8.639680	2 30
6	3 30	.99813	7.720726	10.687	.06105	8.785675	3 30
7	4 40	.99668	7.520498	18.994	.08136	8.910404	4 40
8	6 00	.99452	7.738630	31 388	.10453	9.019235	6 00
9	7 30	.99144	7.932227	49.018	.13053	9.115698	7 30
10	9 10	.98723	8.106221	73.173	.15931	9.202234	9 10
11	11 00	.98163	8.264176	105.270	.19081	9.280599	11 00
12	13 00	.97437	8 408748	146.857	.22495	9.352088	13 00
13	15 10	.96517	8.541968	199.570	.26163	9.417684	15 10
14	17 30	.95372	8.665422	265.186	.30071	9.478142	17 30
15	20 00	.93969	8.780370	345.540	.34202	9.534052	20 00
16	22 40	.92276	8.887829	442.543	.38537	9.585877	22 40
17	25 30	.90259	8.988625	558.153	.43051	9.633984	25 30
18	28 30	.87882	9.083441	694.335	.47716	9.678663	28 30
19	31 40	.85112	9.172846	853.050	.52498	9.720140	31 40
20	35 00	.81915	9.257314	1036.20	.57358	9.758591	35 00

SELECTED SPIRALS FOR A 2° CURVE, GIVING

Δ	$s.$	$n \times c.$	$D_{\delta(n+1)}.$	$D'.$	$d.$
10°	1° 00′	3 × 32	2° 05′ 00″	2° 03′	41.12
10	1 40	4 × 39	2 08 13	2 09	61.04
10	2 30	5 × 43	2 19 33	2 18	73.69
10	3 30	6 × 45	2 35 34	2 33	78.81
10	4 40	7 × 44	3 01 50	2 40	70.47
20	1 00	3 × 33	2 01 13	2 01	45.28
20	1 40	4 × 41	2 01 57	2 02	73.85
20	2 30	5 × 48	2 05 00	2 05	99.99
20	3 30	6 × 50	2 20 00	2 06	109.52
30	1 00	3 × 34	1 57 39	2 01	46.14
30	1 40	4 × 41	2 01 57	2 01	75.16
30	2 30	5 × 49	2 02 27	2 02	109.78
30	3 30	6 × 50	2 20 00	2 02	115.63
30	3 30	6 × 50	2 20 00	2 03	110.90
40	1 00	3 × 35	1 54 17	2 01	46.90
40	1 40	4 × 42	1 59 02	2 01	76.96
40	2 30	.5 × 50	2 00 00	2 01	117.87

EQUAL LENGTHS BY CHORD MEASUREMENT.

½ old line.	½ new line.	Diff.	*x*.	*h*.	*k*.
291.12	291.12	.00	.6516	.040	.061
311.04	311.04	.00	1.702	.187	.110
323.69	323.70	+ .01	3.439	.354	.103
328.81	328.82	+ .01	5.954	.590	.099
320.47	320.50	+ .03	8.955	.897	.100
545.28	545.28	.00	.6719	.122	.182
573.85	573.84	− .01	1.789	.118	.066
599.99	600.00	+ .01	3.839	.527	.137
609.52	609.52	.00	6.616	.554	.084
796.14	796.22	+ .08	.6923	.566	.082
825.16	825.16	.00	1.789	.227	.127
859.78	859.75	− .03	3.919	.377	.096
865.63	865.57	− .06	6.616	.249	.038
860.90	860.98	+ .08	6.616	1.013	.153
1046.90	1047.15	+ .25	.7127	1.222	1.715
1076.96	1077.09	+ .13	1.832	.848	.463
1117.87	1117.77	− .10	3.999	.141	.035

SELECTED SPIRALS FOR A 4° CURVE, GIVING

$\triangle$	$s.$	$n \times c.$	$D_{s\,(n+1)}.$	$D'.$	$d.$
10°	1° 00'	3 × 16	4° 10' 03"	4° 07'	20.22
10	1 40	4 × 19	4 23 13	4 16	29.12
10	2 30	5 × 22	4 32 48	4 39	38.75
10	3 30	6 × 23	5 04 26	5 17	41.37
20	1 40	4 × 20	4 10 03	4 04	34.92
20	2 30	5 × 24	4 10 03	4 09	50.72
20	3 30	6 × 27	4 19 19	4 17	63.69
20	4 40	7 × 30	4 26 44	4 31	78.07
20	6 00	8 × 31	4 50 24	4 46	81.88
20	7 30	9 × 32	5 12 36	5 16	85.40
30	1 40	4 × 20	4 10 03	4 02	35.57
30	2 30	5 × 25	4 00 03	4 04	57.39
30	3 30	6 × 28	4 10 03	4 07	72.37
30	4 40	7 × 32	4 10 03	4 14	93.09
30	6 00	8 × 35	4 17 12	4 23	110.31
30	7 30	9 × 37	4 30 20	4 34	122.20
30	9 10	10 × 38	4 49 33	4 47	126.86
40	2 30	5 × 25	4 00 03	4 02	58.91
40	3 30	6 × 28	4 10 03	4 04	73.75
40	4 40	7 × 32	4 10 03	4 08	94.65
40	6 00	8 × 36	4 10 03	4 12	121.38
40	7 30	9 × 39	4 16 28	4 17	142.86
40	9 10	10 × 41	4 28 21	4 26	154.34
60	2 30	5 × 25	4 00 03	4 01	59.68
60	3 30	6 × 29	4 01 26	4 02	81.04
60	4 40	7 × 32	4 10 03	4 03	99.59
60	6 00	8 × 36	4 10 03	4 05	125.81
60	7 30	9 × 40	4 10 03	4 08	154.42
80	2 30	5 × 25	4 00 03	4 01	58.29
80	3 30	6 × 29	4 01 26	4 01	82.82
80	4 40	7 × 33	4 02 28	4 02	106.99
80	6 00	8 × 37	4 03 17	4 03	135.61
80	7 30	9 × 41	4 03 57	4 05	164.79

V.

½ old line.	½ new line.	Diff.	$x.$	$h.$	$k.$
145.22	145.17	− .05	.3258	.045	.135
154.12	154.13	+ .01	:8290	.080	.100
163.75	163.76	+ .01	1.760	.177	.100
166.37	166.39	+ .02	3.043	.305	.100
284.92	284.92	.00	.8726	.081	.100
300.72	300.72	.00	1.920	.184	.096
313.69	313.75	+ .06	3.573	.375	.105 -
328.07	328.08	+ .01	6.106	.598	.098
332.88	331.92	+ .04	9.191	.910	.092
335.40	335.47	+ .07	13.248	1.310	.099
410.57	410.57	.00	.8726	.137	.157
432.39	432.38	− .01	2.000	.147	.074
447.37	447.35	− .02	3.705	.284	.077 -
468.09	468.09	.00	6.513	.687	.105
485.31	485.32	+ .01	10.377	1.091	.105
497.20	497.23	+ .03	15.319	1.526	.100
501.86	501.95	+ .09	21.240	2.126	.100
558.91	558.88	− .03	2.000	.109	.054
573.75	573.74	− .01	3.705	.361	.097
594.65	594.66	+ .01	6.513	.977	.150
621.38	621.33	− .05	10.673	.973	.091
642.86	642.83	− .03	16.147	1.100	.086
654.34	654.36	+ .02	22.917	2.186	.095
809.68	809.67	− .01	2.000	.180	.090
831.04	831.03	− .01	3.837	.461	.120
849.59	849.52	− .07	6.513	.572	.088
875.81	875.76	− .05	10.673	1.074	.106
904.42	904.36	− .06	16 561	1.718	.104
1058.29	1058.61	+ .32	2.000	.979	.490
1082.82	1082.71	− .11	3.837	.295	.074
1106.99	1107.03	+ .04	6.716	1.000	.149
1135.61	1135.51	− .10	10.970	1.199	.109
1164.79	1164.92	+ .13	16.975	2.440	.144

SELECTED SPIRALS FOR AN 8° CURVE, GIVING

Δ	$s.$	$n \times c.$	$D_{8(n+1)}.$	$D'.$	$d.$
10°	2° 30′	5 × 11	9° 06′ 01″	9° 06′	19.95
20	2 30	5 × 12	8 20 26	8 16	25.71
20	3 30	6 × 14	8 20 26	8 34	34.86
20	4 40	7 × 15	8 53 51	8 54	39.90
20	6 00	8 × 16	9 23 07	9 24	45.52
30	2 30	5 × 12	8 20 26	8 07	26.50
30	3 30	6 × 14	8 20 26	8 14	36.16
30	4 40	7 × 16	8 20 26	8 26	47.01
30	6 00	8 × 17	8 49 55	8 36	53.13
30	7 30	9 × 18	9 16 08	8 46	60.05
30	9 10	10 × 19	9 39 36	9 14	65.70
40	2 30	5 × 12	8 20 26	8 04	26.93
40	3 30	6 × 14	8 20 26	8 08	36.85
40	4 40	7 × 16	8 20 26	8 14	48.25
40	6 00	8 × 18	8 20 26	8 22	61.35
40	7 30	9 × 19	8 46 49	8 30	68.07
40	9 10	10 × 20	9 10 34	8 40	75.01
40	11 00	11 × 21	9 32 03	8 54	82.13
40	13 00	12 × 22	9 51 36	9 14	89.81
60	2 30	5 × 12	8 20 26	8 02	27.30
60	3 30	6 × 14	8 20 26	8 03	38.22
60	4 40	7 × 16	8 20 26	8 06	49.75
60	6 00	8 × 18	8 20 26	8 10	62.87
60	7 30	9 × 20	8 20 26	8 16	77.16
60	9 10	10 × 22	8 20 25	8 24	93.05
60	11 00	11 × 23	8 42 13	8 31	101.08
60	13 00	12 × 25	8 40 28	8 48	118.19
60	15 10	13 × 26	8 58 59	9 02	127.21
60	17 30	14 × 27	9 16 07	9 22	136.45
80	4 40	7 × 17	7 50 57	8 04	57.04
80	6 00	8 × 19	7 54 03	8 06	71.78
80	7 30	9 × 20	8 20 26	8 08½	79.18
80	9 10	10 × 22	8 20 25	8 13	95.23
80	11 00	11 × 24	8 20 25	8 19	112.67
80	13 00	12 × 26	8 20 25	8 28	130.86
80	15 10	13 × 27	8 38 59	8 34	140.88
80	17 30	14 × 28	8 56 13	8 42	150.55

EQUAL LENGTHS BY CHORD MEASUREMENT.

¼ old line.	½ new line.	Diff.	$x.$	$h.$	$k.$
82.45	82.47	+ .02	.8798	.051	.058
150.71	150.72	+ .01	.9598	.051	.053
159.86	159.88	+ .02	1.852	.117	.063
164.90	164.92	+ .02	3.053	.185	.061
170.52	170.55	+ .03	4.744	.221	.047
214.00	214.00	.00	.9598	.049	.051
223.66	223.68	+ .02	1.852	.142	.077
234.51	234.53	+ .02	3.256	.260	.080
240.63	240.65	+ .02	5.040	.325	.065
247.55	247.55	.00	7.452	.287 -	.039
253.20	253.18	— .02	10.620	.590	.056
276.93	276.94	+ .01	.9598	.079	.082
286.85	286.87	+ .02	1.852	.181	.098
298.25	298.24	— .01	3.256	.293	.090
311.35	311.33	— .02	5.337	.330	.062
318.07	318.06	— .01	7.866	.472 -	.060
325.01	325.00	— .01	11.179	.629	.056
332.13	332.12	— .01	15.415	.840	.054
339.81 ·	339.81	.00	20.723	1.024	.049
402.30	402.32	+ .02	.9598	.136	.142
413.22	413.19	— .03	1.852	.083	.045
424.75	424.76	+ .01	3.256	.317	.097
437.87	437.88	+ .01	5.337	.539	.101
452.16	452.18	+ .02	8.280	.863 -	.104
468.05	468.02	— .03	12.297	1.139	.093
476.08	476.09	+ .01	16.883	1.523	.090
493.19	493.18	— .01	23.548	2.160	.092
502.21	502.21	.00	30.817	2.613	.085
511.45	511.45	.00	39.595	3.157	.080
557.04	557.02	— .02	3.460	.366	.106
571.78	571.75	— .03	5.633	.408	.072
579.18	579.18	.00	8.280	.860	.104
595.23	595.25	+ .02	12.297	1.346	.110
612.67	612.70	+ .03	17.617	1.719 ·	.109
630.86	630.90 ··	+ .04	24.490	2.738	.112
640.88	640.88	.00	32.002	3.119	.098
650.55	650.62	+ .07	41.062	3.809	.093

SELECTED SPIRALS FOR A 16° CURVE,

Δ	$s.$	$n \times c.$	$D_{s(n+1)}.$	$D'.$	$d.$
30°	4° 40′	7 × 10	13° 21′ 48″	18° 00′	33.59
40	6 00	8 × 10	15 02 34	17 14	36.14
60	7 30	9 × 10	16 43 31	16 32	38.47
60	9 10	10 × 11	16 43 31	16 48	46.40
60	11 00	11 × 12	16 43 31	17 14	54.62
60	13 00	12 × 12	18 07 48	17 22	54.14
60	15 10	13 × 13	18 01 18	18 10	62.88
60	17 30	14 × 13	19 19 14	18 12	62.85
60	20 00	15 × 14	19 06 05	20 00	72.14
80	7 30	9 × 10	16 43 31	16 16	39.74
80	9 10	10 × 11	16 43 31	16 26	47.49
80	11 00	11 × 12	16 43 31	16 38	56.19
80	13 00	12 × 13	16 43 30	16 56	65.24
80	15 10	13 × 14	16 43 29	17 22	74.72
80	17 30	14 × 14	17 55 44	17 24	75.02
80	20 00	15 × 15	17 50 54	18 06	85.15
80	22 40	16 × 15	18 58 25	18 08	85.18
80	28 30	18 × 16	19 53 20	19 42	95.84

GIVING EQUAL LENGTHS OF ACTUAL ARCS.

½ old line.	½ new line.	Diff.	x.	h.	k.
127.64	127.64	.00	2.035	.388	.191
161.55	161.55	.00	2.965	.430	.145
226.58	226.56	– .02	4.140	.436	.105
234.50	234.45	— .05	6.148	.576	.094
242.73	246.67	— .06	8.808	.860	.099
242.25	242.26	+ .01	11.303	1.093	.097
250.99	250.99	＼ .00	15.409	1.516	.098
250.96	250.97	+ .01	19.064	1.552	.081
260.25	260.25	.00	25.031	2.182	.087
290.55	290.47	— .08	4.140	.328	.305
298.30	298.27	— .03	6.148	.680	.111
307.01	306.96	— .05	8.808	.943	.107
316.06	316.03	— .03	12.245	1.384	.113
325.53	325.54	+ .01	16.594	1.973	.119
325.83	325.81	— .02	20.531	1.939	.094
335.97	335.96	— .01	26.819	2.657	.099
336.00	335.99	— .01	32.276	2.677	.083
346.65	346.66	+ .01	43.221	3.748	.078